Contents

AF808282

Introduction

Software has a secret life.

Nobody has given it much thought.

Software is actually part of *everything*, and this book tries to unravel the *software* in everything – whether in the story of DNA, or how evolution and the brain use software, how intelligence and consciousness works now and can work in the future, the emergence of Artificial Intelligence (AI), the rise of the internet, the world wide web, and the IT Superpowers.

Unravelling the software in everything also eventually leads to considering the future of machine consciousness & machine life, of the world wide web, of the Singularity, (a possible future event where machines overtake humans), and finally, how intelligent, sentient & conscious living machines can share the planet with humans, or end up in conflict with them.

That is a massive area of study, but a rich vein of unexplored connections lay here.

Because in studying this gigantic area, I found that *software* and *life* are intertwined more than many think, and so this book does also focus on *life*, but not exactly as we know it. Software is fundamental to *life*, but it sits in the shadows, waiting to be revealed. Hidden in plain sight.

This book uses *all* that has past, as the basis to *RETHINK* everything about software & *life* & the *future* as we know it.

And along the way, in the middle of research about software & life, I found something. Something really fascinating.

I found a process going on that is almost unnoticed, but powerful and over-arching.

A *vanishing* process. I found so many things vanishing that I realised it was a thing (noun) in itself, not a description (adjective). So, I called it *the vanishing* – the Great Vanishing.

What exactly is *vanishing* became the surprise centrepiece of this book. I hope what I found will surprise you too, even enthral you.

The result is a theory. A theory that says much of who we are, and what we do, and where we are going, is related to a secret life of *software*, and this *Great Vanishing* I discovered.

Every day you can see the world through the eyes of chemists, or physicists, or biologists, or via your own cultural framework. The book invites you to see the world through the eyes of software.

It is a rollercoaster ride into the heart of the modern world of software, but also rethinking and reimagining everything else you thought you knew along the way.

I hope you enjoy the journey.

Jim Ensor,

July 2024

Acknowledgements

I wrote a book 20 years ago. Research 20 years ago was a nightmare unless you were a full time academic. I was never that, I have run a software business for 29 years. So, for me, researching 20 years ago, meant hanging around university libraries, or buying books from second-hand bookshops, or subscribing to journals, or buying expensive new books; always searching for the latest research, the latest thinking.

Coming back to write a second book over the past few years was so different.

I started with 2 tools:

- Kindle Books from Amazon
- Wikipedia

Both are just brilliant, and game changing.

I ended up with literally hundreds of books bought inexpensively from Kindle, and constantly used Wikipedia as a starting point for research. Kindle & Wikipedia naturally let you search by topic, and so pick up everything that might related to your topic of interest. It opened up hundreds of new sources.

The authors' ideas and research I found & referred to, are all carefully footnoted, so you don't see Amazon Kindle or Wikipedia mentioned that much in the actual book, but they were my original searching tool.

Once I had a starting point, I could follow the trail of other researchers and writers, find their work, until finally I ran out of others' research and their good ideas, and had to get down to the hard process of thinking things out for myself, and coming up with new perspectives.

What I liked most of all was how fast you could get to that end point today, compared to 20 years ago.

Thanks also to the brilliant artist and designer Annette, who created the book's front cover. (illustrated & Designed by Annette Ferguson - Imaginette Design).

And finally, thanks to my partner Ju for supporting this endeavour. I could not have finished this without her.

Jim Ensor, July 2024

Publishing Details

© Copyright 2024 Jim Ensor. All rights reserved.

No part of this publication may be reproduced, stored in a retrieved system, or transmitted, in any form or by any means, electronic, mechanical, photocopying, recording or otherwise, without the written prior permission of the author.

ISBN 978-1-7636110-1-6

Publisher details – New Perspectives Publishing Pty Ltd
Melbourne, Victoria, Australia

About the Author –

Jim Ensor is founder of Progenesis Holdings Pty Ltd, a company that for 29 years has been specializing in the development, sales, implementation and support of retail & commercial property & lease management software.

— Part A —

Rethinking Life As We Know It

1. Rethinking Living Things

When we consider the nature of *'living things'*, they can be harder to define than you think. There is no universally accepted way to 'define' whether something is living or not.

Typical characteristics of living things are that they grow, move, breathe, and reproduce. But other important characteristics are nutrition, excretion and sensitivity. These attributes apply to animals or plants. Plants need water and other nutrients, plus sunlight and air. The exact way these characteristics work for each living thing varies, but all feature most of these general requirements mentioned above.

But a 'living thing' is really more than the above. Living things try to preserve their own life, and many actions, like eating for example, reflect that. It's a level of 'awareness' of their existence or needs, that ranges from minimal to higher levels of reasoning and what we call 'consciousness', (which we will return to later).

Living things missing some of the above characteristics or attributes, are usually not considered to be alive. A well-known and quoted example is a rock. A rock does not grow or move or die - so it is therefore not alive.

Viruses

Yet our understanding of what is 'alive' can change. Previously viruses were not considered alive, as by themselves, they cannot reproduce – they need to enter a living cell to do that. They also can't make their own energy, they need cells to do that too.

Recent research by Professor Gustavo Caetano-Anolles and his colleagues at the University of Illinois[1] reveals that viruses and bacteria might share a common ancestor – multiple ancient cells that lived around 3.4 billion years ago. That is not long after life emerged on earth[2].

From then he argues that bacteria evolved more complexity, but viruses evolved (losing genes) to become simpler.

So viruses can be *small*, but carry the key information that allows them to enter a host cell and trick the host cell into making copies of the virus. How small? For example the influenza virus has only 14 protein-coding genes. Many biologists argued that viruses are too small to be considered a life form.

Then around 13 years ago French scientists found a very large virus – mimivirus, (as though it is 'mimicking' a bacteria) with over 1000 genes. And complex. So why isn't it alive? It is hard to argue that it is not alive.

Millions affected by the COVID-19 pandemic from 2020 to 2023, would agree that the virus was alive; it is a matter of lived experience. Nobody now doubts COVID-19 was and is alive.

The Corona Viruses belonged to a large virus family that cause respiratory infections – ranging from a minor cold, up to respiratory diseases that cause death. COVID-19 emerged in Wuhan City in China in December 2019 as a new strain.

But it belonged to the same family as SARS (severe acute respiratory syndrome) and MERS (middle east respiratory syndrome). Its symptoms included fever, coughing, sore throat and shortness of breath.

The ease of spread of the virus (and its mutations) was a main concern during the pandemic, as it challenged the effectiveness of various methods to control its spread, so making it hard to contain and minimise community transmission.

It was certainly alive, and helped change our understanding about what can be considered 'alive'.

The boundaries of what is seen as 'alive' and 'not alive' are shifting.

Extremophiles

Beyond the story of viruses in helping redefine what is 'living'; there is the discovery of *extremophiles*[3].

Extremophiles are creatures that 'live' in extreme environments, so challenging our idea of what is needed, to be defined as being 'alive'.

Thermus aquaticus is a microbe that like at the hot springs in Yellowstone National Park, mainly in Wyoming USA. High temperatures from an underground volcano produce bubbling hot springs far too hot for most creatures.

But in 1969 Thomas D Brock and Hudson Freeze of Indiana University found this heat tolerant microbe that help redefine just how much heat a living thing can stand. And gave us the 'yellow' in the hot springs at Yellowstone.

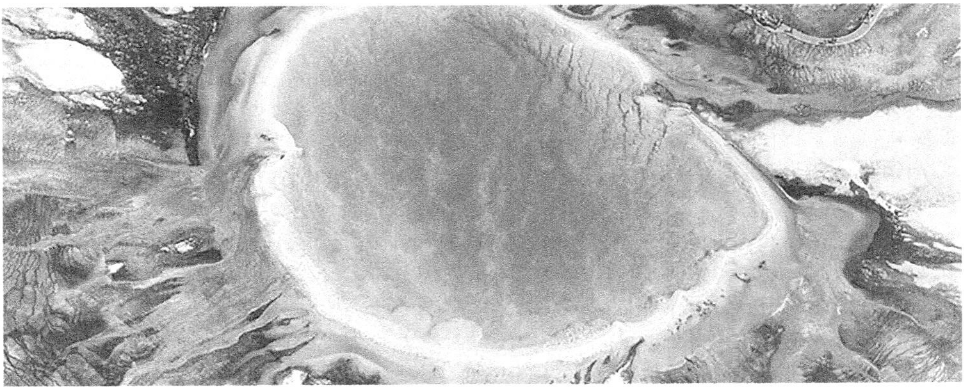

Aerial view of the Grand Prismatic Spring, Yellowstone National Park. Created by Carsten Steger. Wikimedia Commons.

The bright colours are produced by thermophiles. They are a type of extremophile that thrive at relatively high temperatures (41° to 122° C). They are found in many geothermally heated parts of earth, from hot springs (see picture) to deep see vents.

https://en.wikipedia.org/wiki/Extremophile#/media/File:Aerial_image_of_Grand_Prismatic_Spring_(view_from_the_south).jpg

Life emerging in extreme situations – whether in ocean trenches or even surviving on meteors journeying through space, are becoming part of our widening understanding of what 'life' is.

The case of the communities of creatures that live near hydrothermal vents, near fissures on the ocean floor, is another example of life emerging in extreme situations, where heated water rises, often near active volcanos, as per the diagram below:

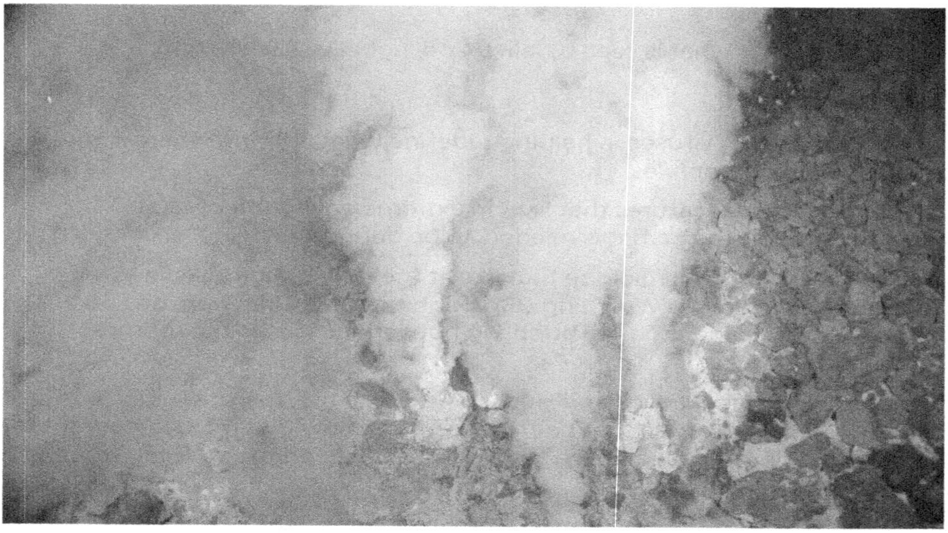

Example of a 'white smoker' venting chemical rich liquid, and supporting some unique life in the Marianas Trench. https://oceanexplorer.noaa.gov/explorations/04fire/logs/hirez/champagne_vent_hirez.jpg

Author – National Oceanic and Atmospheric Administration (NOAA), 2004, public domain

Black or white smoke rising from these vents is filled with dissolved chemicals that form the food chain for larger deep undersea creatures like giant tube worms, snails, crabs and shrimp.

Some believe that life on earth began at these hydrothermal vents, especially the 'white smokers'. That may have been as early as 3.5 billion years ago, with the first evidence of larger, complex multicellular life only appearing around 700 million years ago in the earth's oceans[4].

The worms pictured below (in recent times) can live near vents with extremely high levels of otherwise poisonous and corrosive gas and can grow up to 3 metres long:

See photo of a lush community of vibrant red tube worms growing on a black smoker chimney in the ASHES hydrothermal field. The tube worms, which are hosted in white housings about the diameter of a person's small finger, are intergrown with brown palm worms. (Courtesy of University of Washington, NSF/Ocean Observatories Initiative/Canadian Scientific Submersible Facility) CC BY-SA 2.0 File:Tube worms ASHES hydrothermal field (27260324626).jpg Created: 26 May 2016

https://en.wikipedia.org/wiki/Riftia_pachyptila#/media/File:Tube_worms_ASHES_hydrothermal_field_(27260324626).jpg

The origin and story of *viruses* and *extremophiles* is a tale of survival and adaptation under unusual or extreme conditions.

The definition of 'life' is not as simple as it first seems. It can stretch to include many things, and so viruses and extremophiles are now widely recognised as being *alive*.

But how far will the boundaries of life stretch? The door is open to see what else is, or will come to be, seen as *life*.

Understanding Cells

But leaving aside possibilities on the fringe of biological life, it is the *cell*, and what changes and adapts in *cells*, that is a key to understanding & defining living things as we know them now.

The sheer range of living creatures and survival techniques confirms that so much can be changed in *cell* structure in the struggle to survive and grow.

See below an example of a plant cell, with all its complexity, yet simplicity:

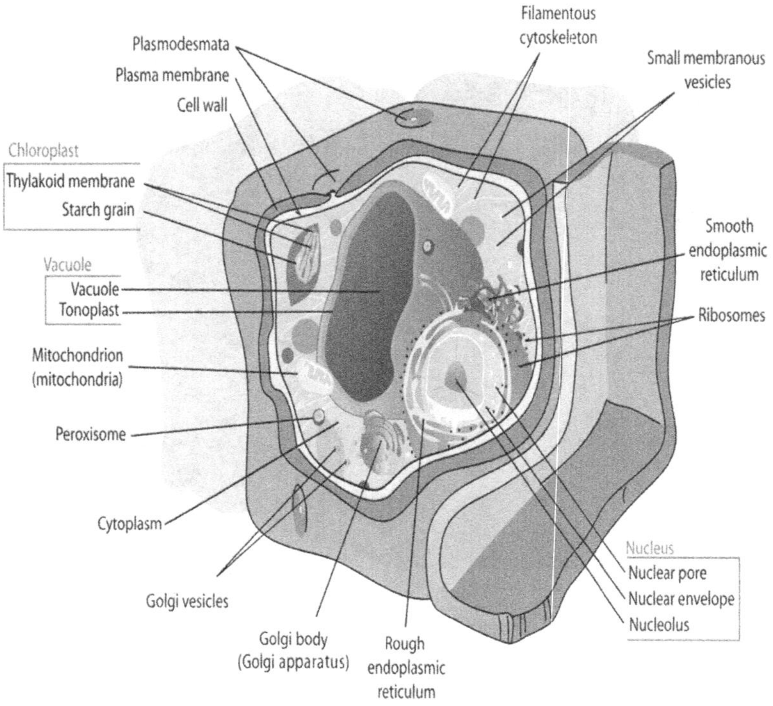

Diagram – structure of a typical plant cell. Creator – LadyofHats using Adobe Illustrator, 2006, Public Domain. Each cell's nucleus has all the cell's DNA – genes structured to perform various functions.

It didn't evolve overnight when you look at the diagram! It is the building block of life, upon which many complex cells combine to give us the richness of life on our planet.

Cell theory is the cornerstone of modern-day cell biology. Cell theory as it has developed and become widely accepted, has several key statements of 'fact' that are the building blocks of *life*.

Many modern texts define *life* in terms of cells, but an intriguing book from 1966 gave one of the most interesting definitions of life. This pioneering book was one of the first to consider the difference between biological and non-biological life. The text is 'Intelligence in the Universe' by Roger A MacGowan and Frederick I Ordway III. More on this book later.

They defined biological life[5] as summarised below:

1. Aggregations of different molecules, primarily complex carbonaceous polymers, with the aggregates containing deoxyribonucleic acid as the genetic agent
2. It reproduces itself
3. It maintains a structural shell or boundary isolating a special internal environment
4. It absorbs or consumes materials or nutrients from the external environment for internal storage or use
5. It utilizes external energy
6. It stores energy in chemical form for later use
7. It releases waste materials into the external environment
8. It senses external environmental conditions and reacts to those conditions
9. It grows by fabricating new structural material from absorbed material or nutrients
10. It eventually dies, the cessation of all normal chemical processes then decomposition.[6]

But in today's world, we can start to understand *life* as having biological cells at its core, as cells are the simplest unit of matter that is seen as 'living'. Cells are the foundation of *biological life*. Two main kinds of cells are eukaryotic (has a true nucleus) and prokaryotic (no true nucleus).

All living organisms consist or 1 cell or more. Even a single cell organism is complex in some ways.

Cells are the building block of biological life that allow coordinated actions. And because coordinated actions exist in living things, it follows logically that living things are *organised.*

Instruction Sets

High organisation needs goals and a way of achieving them – in short, some instruction set.

The importance of a living thing being partly defined by having an 'instruction set' opens up all sorts of possibilities for defining what is 'living'.

But just what kind of 'instruction set' is it?

In a word – DNA or Deoxyribonucleic acid.

DNA

DNA was originally discovered by Friedrich Miescher, who identified DNA as a distinct molecule in 1869. But it was in 1953 that the 'double helix' or 'twisted ladder' structure of the DNA molecule was discovered by James Watson and Francis Crick.

It was a massive discovery and pivotal point in understanding life as we know it, the key that unlocked the human genetic code. The Nobel prize for this discovery of the molecular structure of DNA was made by Watson, Crick and Wilkins. It will always remain an amazing achievement. But a shadow still hangs over the discovery.

Mention needs to be made of the role of Rosalind Franklin in the Watson, Crick & Wilkins DNA discovery. It is one of those fascinating tales of academic intrigue.

Frankin's graduate student took an X-Ray now famously called X-ray Photo 51, that showed the double helix structure, and this found its way to Watson without Rosalind's permission (it went to the assistant director of the lab, with her permission, who gave it to Watson). Rosalind seems a generous person, and probably didn't mind.

Yet using someone else's discoveries to confirm your own research, without formal permission or acknowledgement, is not actually fair to the person doing that original work, if that is what happened. See a picture below of Rosalind while working:

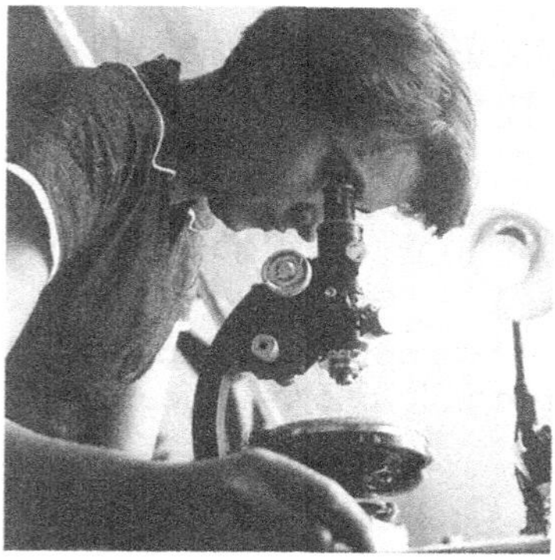

Rosalind Franklin with microscope in 1955. From the personal collection of Jenifer Glynn.

Author - MRC Laboratory of Molecular Biology

The X-Ray photograph 51 (below) has been called the most important photograph in history, for the doors it unlocked.

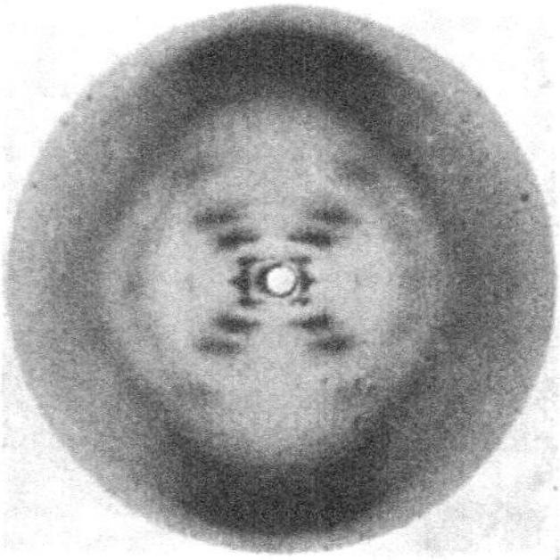

Photo 51, Wikipedia, 30 Dec 2012, en.wikipedia.org/wiki/File:Photo 51 x x-ray diffraction_image.jpg

The most important photograph in history.

And equally importantly, it was discovered recently that a page by Franklin on her work went into a Medical Research Council (MRC) report, that was given to Crick by his supervisor Max Perutz without her knowledge, which was vital to confirming their (Watson & Crick's) model.

So there was some 'boys club' aspect to this story, where the 'boys' were passing each other information, racing to the finish line, and helping themselves to the credit, and of course the Nobel Prize for Physiology or Medicine in 1962.

Whereas Rosalind was more a dedicated independent female researcher rather than a skilled political operator; and saw the research as something they were all working on in some kind of shared search for knowledge. She was definitely the victim of the boy's club. Or maybe worse than that.

Rosalind developed cancer, and died in 1958. She was friends with Crick and his wife, so she never felt badly treated. But would it have been so hard to recognise Roselind Franklin's role in the DNA discovery at the time, and share some of the fame?

The discovery of the double helix structure led to the emergence of modern molecular biology, aiming to explore this whole DNA area. A whole world of human genetics had opened up.

DNA is a chemical code used as instructions to construct living things, code that scientists have decoded and are starting to be able to manipulate. Code with its own alphabet.

To see its double helix shape is an amazing sight -

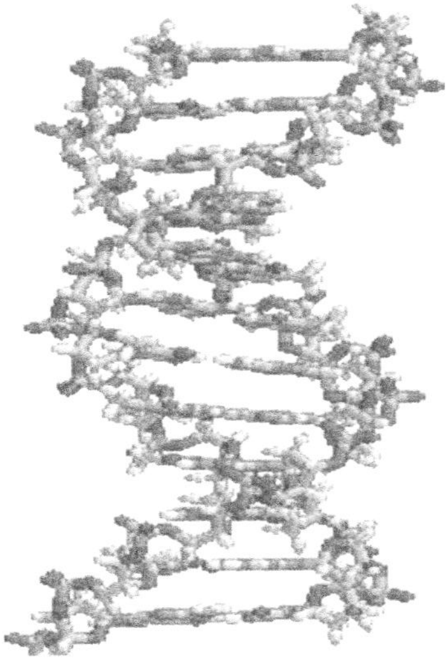

By Brian0918 - Single frame of Image: ADN animation.gif, created with data from Edwards K, Brown D, Spink N, Skelly J, Neidle S (1992). "Molecular structure of the B-DNA dodecamer d(CGCAAATTTGCG)2. An examination of propeller twist and minor-groove water structure at 2.2 A resolution". J. Mol. Biol. 226 (4): 1161–73. PMID 1518049., Public Domain,

https://commons.wikimedia.org/w/index.php?curid=2052637

The discovery of the double helix shape was the trigger of an explosion in DNA research and discovery, which is still happening today.

DNA Code

To understand DNA it is best to see it in the context of an analogy.

Author Beth Skwarecki[7] made the excellent analogy of comparing the structure of cells & DNA to *cooking in the kitchen.* It is a simple way to pick up the gist of technical terms and get your bearings.

Genetics	Cooking in the Kitchen
DNA – Instructions in chemical code to build and maintain every part of the body	Recipe for a meal
Cells - contain all of our DNA. They read the DNA code and carry out the instructions eg. skin cells use the genes to build skin cell related things. We keep 2 copies of our DNA in each cell – the from mum and from dad. We have around 37 trillion cells.	The kitchen where recipes are made. Different recipes are used for different occasions – eg. roast dinner for a special occasion, or salad for a light meal.
Gene - short set of instructions in the DNA containing the instructions to make a tiny piece of who we are eg. brown hair if the genes encode that pigment	Different recipes. Each recipe can vary slightly. So each meal ends up a little bit different.
Genome - all the information carried in our DNA. All the information carried in our DNA would fill 12 thousand books.	Deluxe cookbook featuring lots of extra notes on how to plan a dinner party or menus for a week's dinners. Or could be called a 'cookbook encyclopedia'
Non-coding DNA – scraps of DNA	Like a cookbook, where there might be many versions of one dish, but only one is good to make
Chromosomes – one string of DNA split into 23 pieces	Like a recipe collection in 23 large volumes
Molecule – made of millions of atoms, DNA is a huge molecule. Molecules are the smallest piece of a compound, eg. water or DNA.	Imagine a glass of water, with the smallest item in the glass is a single H2O molecule

Genetics	Cooking in the Kitchen
Atoms – the smallest possible piece of an element, the building blocks of molecules. Elements in the periodic table are the flavours that atoms come in eg. carbon, hydrogen etc	Even smaller than a molecule of water
Atoms in DNA - most molecules in our cells can be made with just 6 atoms -carbon, hydrogen, oxygen, nitrogen, phosphorus, sulphur. The first 5 are needed to build DNA	Materials used to create the ingredients
Building Blocks – DNA has 4 different types of building blocks, and these pieces come together in a structure called a double helix ('helix' from the Greek word for a spiral).	What the ingredients looks like
DNA 'code' – is spelled in an alphabet of 4 letters - A,T, G, C or • Adenine • Thymine • Guanine • Cytosine They are called nitrogenous bases. Each represents a different kind of nucleotide (the building blocks of DNA)	How to build a recipe

This discovery that DNA had its own code was truly remarkable.

More than that, the simplicity of the code is startling. There are only 4 letters in the code library:

A (adenine)

T (thymine)

G (guanine)

C (cytosine).

But this coding writes the 'instruction set' used to create what we call *'genomes'* of living things – all of its hereditary information as coded in its DNA. The 4 letters of code create genes underpinning most living things on planet earth.

A code with only 4 letters (A,T,G,C) is able to create thousands or different proteins located in our cells. The process is called the 'modular paradigm of molecular biology' and the most elegant way of describing the process was put by Nessa Carey[8] , who made the analogy of using *'Lego'.*

'Lego' features only a limited number of blocks and designs, but these blocks can be used to create models of everything, from houses, animals, trees, ships etc. Nessa notes that the 'proteins' are like these Lego creations – they are made up of 20 amino acids that can be joined together in any combination.[9]

The 4 bases in DNA 'encode' the 20 amino acids when the cell 'machinery' reads the DNA in blocks of 3 base-pairs at a time (called a codon). So it could be a codon consisting of GCG or AAG for instance, which could mean up to 64 possible codon combinations encoding the 20 amino acids. And some acids are coded by more than one codon. The DNA act as a 'script' for producing proteins via molecule called messenger RNA.[10]

And any little changes to the code make a real difference, just as with ordinary computer software coding.

Changes (eg. an insertion or deletion) to the 'base' DNA code are called a 'frameshift' mutation, that will alter the sequence of code and could mean a drastic change, even if it is a single letter altered. An example Skwarecki uses[11] refers to the code AUG AUG AUG AUG AUG. If the second 'U' is deleted, then it becomes AUG AGA UGA UGA UGA UG.

The result of that single change in this case is that a protein, which is being made in a cell, is signalled to stop work.

This *is* code that, when put together in a certain way, runs a program. It is a kind of *software* code, *biological software* - with real impact. The output can be *biological changes, or even biological life.* And how does understanding this code, help explain how things change?

CRUD

We can say that DNA *coding*, as discussed up to this point; is starting to sound very similar to other software *coding* written to be executed by computers.

We know about the big point of difference. DNA code, when executed in biological software, can create an organic life form. Whereas typical software code for computers (up to now) just typically creates a software program or package, that when executed might help you use a spreadsheet, or buy a ticket online, or drive a bus.

The proof of the similarity has to lie in the detail. Is there any difference between what a biological software developer and a computer software developer can do?

In particular, can a biological software (DNA) programmer, like a typical software (computer) programmer, change code by editing it, or copying it, or deleting it? The answer lies in a single word – CRUD.

The computing acronym CRUD is used to describe 4 key software functions in areas ranging from managing the database, to handling the user interface.

CRUD stands for *create, read, update and delete*. There are some alternative acronyms like BREAD (browse, read, edit, add, delete) or CRAP (create, replicate, append, process) that describe the same functions. CRAP is good from the point of view of mentioning *replicate*. Replication is such a powerful ability, and to a lesser or greater extent, allowing for variations, we are all products of this process. But let's stay with CRUD, it is widely recognised.

These are the functions that make computer software workable and useable, as you need all 4 parts of CRUD to manipulate data.

In the world of computing, once you enter (Create) a contact record, for instance, of John Smith, you need to be able to see his details, (Read), to change the details like his phone number (Update), and finally remove the data when no longer needed (Delete).

The term 'CRUD' is widely credited to James Martin in his 1983 book 'Managing the Data-base Environment'[12] , and is a hallmark of typical coding activity and successful software functionality.

Does DNA allow CRUD functions? That would be the 'test' of whether you can really 'code' DNA, as opposed DNA being made of 'code'.

And the answer is 'Yes'. In summary:

C – CREATE - DNA is at the point where it can be created in test tubes.

In 2003 a sheep named 'Dolly' was cloned Dolly was cloned at the Roslin Institute in Edinburgh Scotland, based on a cell from a mammary gland – thus the name 'Dolly' (after Dolly Parton's glands). The birth of a normal clone proved the idea that any cell could form the foundation to rebuild a whole new individual. These are the early days of a long and controversial journey.

R – READ - DNA can definitely be read, we know the letters that form its code, and we have learned how to read it.

U – UPDATE - DNA can updated. (In nature, a typical insertion is a virus, like COVID-19, where the virus inserts its DNA into a person's DNA). 'Update' can mean 'copy' or 'duplicate', of course, a big part of DNA's function.

However editing the DNA of an organic & living entity is naturally MUCH more complex (and risky in many cases) than changing computer code in a software program of a computer.

Yet a whole new world of editing DNA is now emerging. CRISPR-Cas9 is seen as the most advanced tool to edit genes within organisms. It is used for biological research, treatment of diseases, and helping in the development of biotechnology products. The editing process is one of the frontiers of DNA research.

The first U.S. trial of CRISPR in humans has begun, as reported by Popular Science in April 2019[13]. Two patients were taking immune cells from the patients' own bodies and editing them with CRISPR, then putting them back in, to help treat a difficult form of cancer, at a University of Pennsylvania study. Risks abound.

What if they edit the wrong gene, or if the editing makes no difference, or even causes harm? But it is just the early stages.

In 2020 a new kind of 'base editor' was developed to gene-edit Mitochondria for the first time, potentially making it possible to treat disorders relating to faulty Mitochondria[14]. This is a fast-moving area, with continual new progress.

D – DELETE - DNA can be deleted. (In nature where DNA is missing a section, it is called a 'deletion'. That deletion might be only one missing 'letter', or it could be a whole gene missing). Once again, harder in the case of organic creatures.

But in 2017, scientists created a way to 'delete' DNA in living cells. They developed what they call 'molecular scissors' to literally snip away disease-causing DNA. They used a software tool called DECKO from Johnson Lab to help.[15]

CRUD in computer programming are the four basic functions of 'persistent' storage. That is, 'persistence' is the state that outlives the process that created it. At the finish there is a result, an output that the process produced.

These 4 CRUD functions are the key definers of ability to program, as you can create, see, change then remove.

Another Way of Defining 'Life'

DNA, by allowing CRUD, fits into this description of typical coding activity that helps define software. DNA is not just made of code. You can really 'code' DNA.

The core chemical combinations are the raw material of life. And DNA code is programming code.

The role of *programming* in 'life' became even clearer on May 2010, when J Craig Venter led a team of scientists that, for the first time, created what we now call 'synthetic life'.

They *coded* a fully synthetic bacterial chromosome and loaded it into emptied host bacterial cells. Subsequently the host cells could grow and replicate, a key marker of a living thing – a 'living' artificial cell.

Or, as he describes it, in May 2010 his team synthesized an entire bacterial chromosome from computer code and four bottles of chemicals, then booted up the chromosome in a cell to create the first synthetic organism.[16]

Great description of his tools of trade – computer code and 4 bottles of chemicals, and an amazing story!

The key takeaway of this process of creating synthetic life, is that life is programmatical at its core, and that understanding life as *'biological software'* allows us to rethink and reimagine this whole area.

It is another way of seeing 'life'. The new equation is:

Chemicals (Chem) + *biological* software (BS) = *biological* software life (BSL).

Chemists and physicists have been 'crab-walking' *towards* this definition for a few years, and maybe are not there yet.

It seems too simple.

The link between *chemistry* and *biology* has been looked at for years, with important work from evolution scientists like Addy Pross (Professor of Chemistry at the Ben Gurion Uni) bringing it into the spotlight, especially with his 2012 work 'What is Life? How Chemistry becomes Biology.'[17]

Addy Pross defined life as a network of chemical reactions, or the chemical consequences that derive from the power of exponential growth operating on certain replicating chemical systems.[18] The highlighting of the role of chemical reactions in life was a very important contribution. But software was not seen as a part of the equation or definition of *life*.

Similarly, in 2016 an important paper[19] 'The informational architecture of the cell' from respected author and physicist Professor Paul Davies and co-authors Dr Sara Imari Walker & Assistant Research Professor Hyunju Kim (all from Arizona State University) proposed that the building blocks of life may be *information* – not chemicals or cells as we know it.

It was a radical idea in 2016, (and probably still is), that could be a fundamental building block of life, where information is *organised to direct causal flows*.[20] But the critique of this, is the obvious question, of how exactly does information *cause* things to happen? (Same issue as with a chemicals-based definition). The paper proposed that somehow information processing may *intervene* to cause activity. But it is not totally clear how *information* alone can make things happen.

It all tantalisingly points to *software*, as software is designed to manage, manipulate, change and store information, but the word 'software' does not appear in the paper. If it is software, then it would be a kind of software that we can barely understand at this point of time, in 2012, or 2016 or even today.

Paul Davies also released his book *'The Demon in the Machine'* in 2016. He always writes well-argued books full of great insights, and this book is no exception.[21] The book sharpened the focus on what he saw as separating life from non-life – *information*.[22]

The definition of *life* had swung to be seen as, in the words of biophysicist Eric Smith, 'a chemical system in which the flow and storage of energy are related to the flow and storage of information'.[23] But just how did information become life, was still the core question here?

Paul Davies was definitely thinking about software as being more central to life, when he made the observation in passing that biological information is the software of life.[24] And he did see DNA as *software* when the instructions for building a protein are read out from DNA, and the protein is made by a ribosome.[25] He saw *information* as the bridge between physics and biology.[26]

And in bringing *information* to centre stage, he and others like him helped change the debate about life. *Software* was slowly emerging as part of the equation of life, but nobody knew exactly how the 'physics of information' applied to living systems.[27] But it still does not sit at centre-stage.

However, the foundation stones for the idea that *software* is the magic ingredient turning non-life into life were being laid.

Summary

We originally looked at how the 'stricter' definition of *life* used to exclude viruses and extremophiles, but a rethinking in the broader scientific community brought them recognition as being 'alive'. The boundaries of life were not as fixed as appeared on first glance.

We then wandered through the work of chemists and physicists to define life; and found that the debate was still underestimating *software* as a key element.

Understanding the role of *software* is the focus of this book, and understanding its 'secret life', things that remain hidden or unexplored about software.

That search for understanding starts with biological software, a special kind of software that we are a long way from understanding in its full complexity, but something that we are beginning to seriously explore. It is software manipulating information in ways we cannot yet fully understand or replicate in all its complexity.

So, the boundaries of life, now understood as *biological software,* stretch to include *viruses* and *extremophiles.* Can the boundary shift even more?

Since software is central to biological life; and is obviously at the same time right at the heart of computer-based software, then possible future computer software-based non-biological life might be possible. In which case all *life* is talking the same language, and we may be closer to a unitary theory of life than we think.

It opens the door in the future for other things to be seen as 'alive' if they meet certain criteria, yet to be understood or defined.

In particular, can an artificially intelligent (AI) piece of software, written in another code set, also have the potential to become alive? It's all code, after all. The arrival of AI will start to stretch the definition of what is living even further than we can imagine now.

This is the frontier of *non-biological life*, and all its mystery. We will return to this topic later in the book (Chapter 12).

As our understanding shifts and widens on what is living, the distinction between biological life and non-biological life is *vanishing*. It's all code.

2. Rethinking DNA

Yet if software was so important to life, why did DNA never start off as being seen and defined as 'software'?

Why DNA *'code'* is not often described as 'software,' is based in history not logic. Where occasionally DNA was referred to as 'software' in the past, it was in terms of an analogy, or point of comparison. What is being said here is that DNA is 'code', but it is also much more than that.

'Software' is classically defined as 'programs, etc. used to run a computer'.[28] And a 'program' is classically defined as 'a set of instructions in code that control the operations or functions of a computer'.[29]

So in popular culture, the reason why DNA code is not seen as 'software', is that software code, while based on a coded instruction set, is executed on a computer, not a 'living' thing, according to its original 'popular' definition.

While on the other hand, a biological instruction set, is a set of code in biological form, and executed on a living organism.

Why DNA code is not regarded as 'software' mainly comes down to a *definitional* and historical question as much as anything. Is the instruction set that defines the way a living thing appears & behaves - a *code* set?

'Code', in computer terms, is the term (noun) to describe instruction statements in a computer 'language', that when put together are a program to achieve a certain result. There can be source code, the original statements. And there can be object code, which is the source code consolidated or compiled in a special format for actually running the software program.

The same applies to DNA. It is biological code, and has its own 'language'. The code has its own methods & style, which we will examine in more detail later. In particular, we will look at how the code is activated or not.

The more we understand about DNA, the more we recognise that DNA does actually behave like any other code set, including some subtleties only discovered recently.

The output of the DNA code is living things, while now the output of traditional software programs running on computers is typically only a file or a printed report. But the line is starting to be blurred, as we shall discover in more detail later.

Biological Computing

If any more evidence was needed of the crossover and convergence in this whole area, in 2017, music was encoded in DNA, and then retrieved for the first time – in perfect condition.[30]

The songs recorded were 'Smoke on the Water' (Pink Floyd) and 'Tutu' (Miles Davis). They were saved as a DNA file by a biotech firm called Twist Bioscience, who collaborated with the University of Washington, Microsoft, and the Montreux Jazz Digital Project.

The process started with a digital file, which was then converted from '0's and '1's (Binary code), into 'As','Cs','Ts','Gs' (DNA code). So a 10 could become a C, and an 11 could become a T etc. The result of this conversion was then turned into synthetic segments of DNA grouped into numbered sequences of around 12 bytes, that help locate the data within the DNA file. This is then checked for accuracy, and the job is complete and the music is stored. Amazing!

The final job is to decode the DNA code (A,C,T,G) and turn it back into Binary code (0,1). Once decoded, the file can be opened and the music played. The 2 songs were played at the Montreux Jazz Festival in Switzerland, perfectly accurate & fully preserved. UNESCO added these first DNA files to their 'Memory of the World Archive', in recognition of their cultural importance, as a milestone of a kind.

Ability to encode and decode between 2 coding languages was also of great significance that needs to be recognised now as a real milestone, telling us that the door is opening to a whole new era of understanding, using and, in some situations, changing DNA.

This type of coding falls into a new category on the pioneering edge of which is called biological computing (or biocomputing). It is also a story about the power of DNA storage capability.

In another example, biologists are starting to have combinations of cells carry out computations, acting in effect as biological computers – replacing microchips with cells. All cells have pathways that respond to incoming signals by switching genes on and off, and make the building of biological computers possible. They are compact and energy efficient.

Work in this area has been going on since the 1990's, and the ability to edit genes has allowed in one case study, (a group at the Swiss Federal Institute of Technology in Zurich in 2019) for a kind of CPU (central processing unit) to be inserted into a cell, (using gene editing) to do calculations, for instance. The processing speed is far slower and setup more complex than with using a microchip, so limiting commercial possibilities.

A team in China in 2023 were working with Associate Professor Fei Wang at Shanghai Jiao Tony University on a liquid computer that can use strands of DNA to run over 100 billion different single programs. They are making circuits like on a computer chip, but with DNA molecules acting as wires and instructing wires to configure in certain ways. Their idea is to build tools to help diagnose diseases.[31]

Biological computing is far from mainstream computing at the moment, as nothing developed so far is very useful. It is really a process of re-engineering existing life then building new life. But, as investment continues, this could change. The reason to point to it here is to highlight the emerging ability to *program* DNA.

In another sign of the way things are moving, there is now a new field of study for the development of software tools to understand biological data like DNA - *bioinformatics*. Where data sets get so large and complicated, things move past what an Excel spreadsheet can manage, and better software methods and functionality is needed to manipulate biological data.

A Matter of History

Beyond the evidence above of DNA behaving just like another piece of software, there is the *historical* context to consider.

If we were finding out for the very first time right NOW, that living things had coded instruction sets, we would be calling it a *kind* of software. It's the most accurate way to see it.

Many things are based on *code*, yet people don't see them as software. That's because software coding, as we know it, really emerged after World War II, and only started to take off in the 1950's and 1960's.

DNA hasn't been seen as *software* mainly because although its structure was discovered in 1953, it was already discovered and identified as a distinct molecule by Miescher in 1869 – almost 100 years before the dawn of the modern computer age in World War II, and well before we understand what modern 'software' is.

Turing, one of the pioneers of modern computing, died in 1954, and thus the legacy of his work was not understood at the time, and modern computing was just emerging at the time of his untimely death. (More on this later).

So it's no surprise that DNA code, and all that followed the 1953 discovery of the DNA structure, became part of biology not computing. Because nobody knew exactly what it was then. DNA couldn't be defined in terms of something that barely existed – like a software program.

The key conclusion coming of the historical timing (ie. DNA defined before modern definitions of software) is that the 'classic' and original definition of 'software' - as code relating to computers – is now out of date, as is the understanding of DNA.

'Software' can now be redefined as ***'a coded instruction set that is either biological or machine based'***. This can be run on a computer or biological entity. 'Computer' digital code is usually a series of 0's and 1's (but future computers may use another basis for coding). 'Biological' code is based on DNA.

The original definition of software as being 'tied' to a computer, is out of date. DNA is biological software. It is based on the CRUD foundations of an instruction set, just the same as computer software.

Let's consider how computers work for a moment. Behind each computer is a digital instruction set to program instructions.

The term 'digital' means using a system that performs operations or carries out instructions, based on the 1 or 0 digits of the binary number system, that is used by the vast majority of computers.

Basically any letter or number or combination of these 1's and 0's can be programmed to create anything we like – if we know how. It is different in style from the earlier discussion of DNA, where (biological software) programs using a much smaller (but still extremely powerful) toolset.

We can thus distinguish *biological software* based on DNA code (A,T,G,C), from *machine based computer software* based on digital code (0,1). Or can we?

The Hidden Code

If *software* can be defined as not JUST code written for computers, but can include some other types of coded instruction sets, like biological code, what does that mean?

It begins a search for anything that is based on code, yet people don't see it as software – the *'hidden'* code that is really software by another name.

The search start with things that people already often refer to as *'code'*. The implication here is that many 'instruction sets' – wherever they are – have been 'misclassified' as something else – instead of being seen as a form of 'software'.

And also keep this in mind – software can be a LOT more, than simple or complex instruction sets. Many types of code now can be built to emerge or adapt or evolve, or take different shapes, based on different inputs.

There are many suspects in the initial search for *'hidden'* software, or software that is 'mistakenly' called something else.

The search begins with the origin of *science*, and its many branches.

What happened in human history is that *'science'* developed for centuries in differing fields of endeavour – such as maths, physics, chemistry, and then developed into a whole bunch of new disciplines like in the diagram below:

Hierarchy of the Sciences -

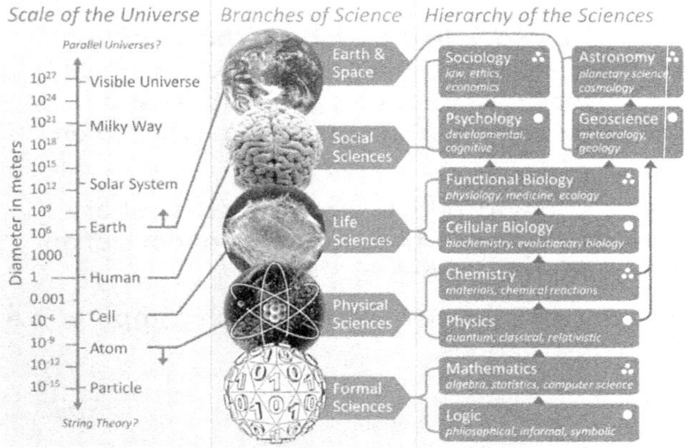

Attribute to - By Efbrazil - Own work, CC BY-SA 3.0, https://Commons.wikimedia.org/w/index.php?curid=25567651

So when we look at the 'Hierarchy of the Sciences' diagram above, the following areas are largely composed of software 'code'[1] –

Original Branch of Science	Use of Code	Redefinition of Branch
Logic	The basis of all code	Stays as is
Mathematics	The basis of all code	Stays as is
Physics	The laws of the universe are a set of rules, read this as programming rules	Physics Software
Chemistry	The period table is a set of codes	Chemical Software
Cellular Biology	Central here is DNA otherwise known as deoxyribonucleic acid, an area built on code from simple to complex	Biological Software
Functional Biology	Management of complex systems	Biological Software
Astronomy	Complex maths & formulas are at the heart of this area	Space Software
Geoscience	Complex maths & formulas are at the heart of this area	Earth Software
Sociology	Higher level theories	Stays as is
Psychology	Higher level theories	Stays as is
Philosophy	Higher level theories	Stays as is

In short, with the benefit of hindsight, we would have SOFTWARE as the highest level of science, and most of the sciences fitting into that framework.

Applying a broader definition of 'software' to many fields of intellectual endeavour is resetting science to the 21st century.

Sure, lots of different things happen in each scientific field, but it is time to reboot the whole discussion, given the fast movement in software that is transforming all these fields.

Many scientific disciplines emerged centuries *before* the emergence of the modern-day software industry that rose after World War II.

Why would anyone consider <u>science</u> done prior to modern day software, as 'instruction sets' or 'code'?

The answer is that no one (at the time *prior* to computers) did, as nobody knew the reality – that CODE or instruction sets can form the basis of *software*. And at a time when software was not 'defined' as existing; code as developed, became absorbed into different branches of science.

Nobody has gone back in time and said 'Hang on, that 'code' is really a piece of what we would now redefine today as software', and some of those old definitions need to *vanish*.

3. Rethinking Evolution

Evolution Theory

What does rethinking the role of code in DNA mean for the way we understand evolution?

To begin with, it is hard to understand evolution, without starting with the work of Charles Darwin.

Up until the mid-1800's, the prevailing view about evolution was that it did not happen, and that things were created in their current form by God.

The Darwinian view is essentially that, based on a distant common ancestor, we evolved over time into modern human beings. From an ape-like creature to Homo sapiens, via 'natural selection'.

By 'common ancestor', it means that we today are just one branch of a tree, the one that survived and thrived, while others coming from the same gene pool either vanished, or remained in the jungles swinging from the trees, like our closely related cousins the monkeys, or the remaining gorillas living in the mist of some beautiful African mountains.

Darwin's intellectual journey really began in the hot and rugged Galapagos Islands off the coast of South America in 1835. There he found thrushes, finches, wrens and doves unfamiliar with humans.[32]

He found strange patterns of tortoises and mockingbirds that were the similar, but a bit different.ii He also found *different species* of some birds and tortoises, on different[33] islands.

Charles Darwin

The variety of finches struck Darwin. There were finches with small beaks for catching insects, adjusted ones to feed on flowers or fruits, or strong thick ones for cracking nuts and seeds.[34] Edward J Larson noted in 'Workshop – God and Science on the Galapagos Islands', that it was these wide differences in the finches were a key inspiration for his theory about species changing through accumulated variations.[35]

1. Geospiza magnirostris.
2. Geospiza fortis.
3. Geospiza parvula.
4. Certhidea olivasea.

Darwin later realized that his observations provided the evidence he needed for a more general theory of evolution, which was based on the view that certain different species of birds on each island had *evolved* from the same distant ancestor in some time past, when those original birds arrived at the islands. Then as time went by, and they settled on different islands, they had *adapted* to different situations on each unique island, as survival demanded.

In 'On the Origin of the Species', published in 1859, Darwin proposed his theory of evolution based on 2 key factors – variation and selection. 'Variation' means that the children are not exact copies of the parents – they vary slightly. And 'selection' means that there is a winnowing out of those who for some reason are less able to survive and reproduce.[36]

For a bird, if having a long beak helped it to survive, then it would pass this feature along to its offspring. If we started with a population of birds with a variety of beaks, and if this advantage of a longer beak was great, then in the next generation, the average length of beak would be greater, as those with the shorter beaks would miss out in the struggle for food, and leave less descendants.[37]

Important related developments were in heredity, especially the work of Mendel, which were only appreciated much later. Mendel was born and mainly lived in Moravia, now in the Czech Republic not far from Prague. Mendel, the monk and teacher, experimented in the period 1854-68 with the breeding of peas, and developed a model to understand how individual characteristics are inherited.

He found that characteristics could be passed onto the next generation; for instance, whether the peas were taller, or smoother, or had different colours. But the interesting thing was that a characteristic passed on, could appear, or may be suppressed, then could appear in a later generation. This finding gave us an understanding of the complexity of inheritance, and the startling process of heredity.

So evolution, as we understood it in the later 20th century, was at last understood to be powered by *variation, selection* and also *inheritance.*[38] It became the scientific orthodoxy about evolution, until discoveries in the early 21st century.

Eventually this new knowledge of the importance of small mutations, from inheritance, would be added into Darwinism to help produce modern neo-Darwinism, as the new orthodoxy by the mid-20th Century

So Darwin's breakthrough explanation, in the 19th century, saw the *vanishing* of many myths of evolution; and laid the foundation for new fields of study like sociobiology and evolutionary psychology (and others) in the years ahead.

Sociobiology aims to explain social behaviour in terms of evolution.[39] It is closely related to evolutionary psychology. Evolutionary psychologists see human behaviour from an evolutionary perspective, looking for *psychological adaptions*[40] that evolved over time to solve any physical, social, economic or environmental problems.

Modern humans – *Homo sapiens* – can be seen as products of this variation, selection, and inheritance process, and there was a growing willingness to explore the implications of this evolution process in new fields of study.

The Neanderthal Story

While those new fields of study were opening up, new information about Homo sapien DNA started to emerge. In the past few decades, it was found that some DNA had been 'inherited' from our 'cousins' the Neanderthals.

More detailed studies of Neanderthal DNA came out of 'The Neanderthal Genome Project', started in 2006, and initiated by a biotech company called 454 Life Sciences but coordinated by the Max Planck Institute for Evolutionary Anthropology.

The Max Planck Institute's team led by Svante Pääbo completed the first draft of the Neanderthal genome (ie. the sum total of their DNA) by February 2009.

Svante Pääbo director of the Department of Genetics at the Max Planck Institute for Evolutionary Anthropology and head of its Neanderthal genome project.

Permission details - *This file was published in a* ***Public Library of Science*** *journal.* ***Their website states*** *that the content of all PLOS journals is published under the Creative Commons Attribution 4.0 license (or its previous version depending on the publication date), unless indicated otherwise.*

Subsequent research by them and important work by others updated this initial work and added more detail based on additional Neanderthal bone fragments found in Siberia, Croatia, Spain, Germany and Russia.

The likely genetic contribution of early 'homo' like Neanderthals was around 1 – 4% of homo sapiens DNA.

Neanderthals and Homo sapiens split into 2 separate lines around 370,000 years ago. Then anatomically-modern humans evolved from 'archaic' humans around 200,000 – 300,000 years ago in Africa and then migrated to Europe between 40,000-100,000 (approx.) years ago, *replacing* the Neanderthals and other archaic humans, then repeated the pattern over the rest of Asia and South East Asia, finally ending up in Australia even as early as 65,000 years ago,[41] as per the diagram below. All other lines of *humans* slowly or rapidly became extinct.

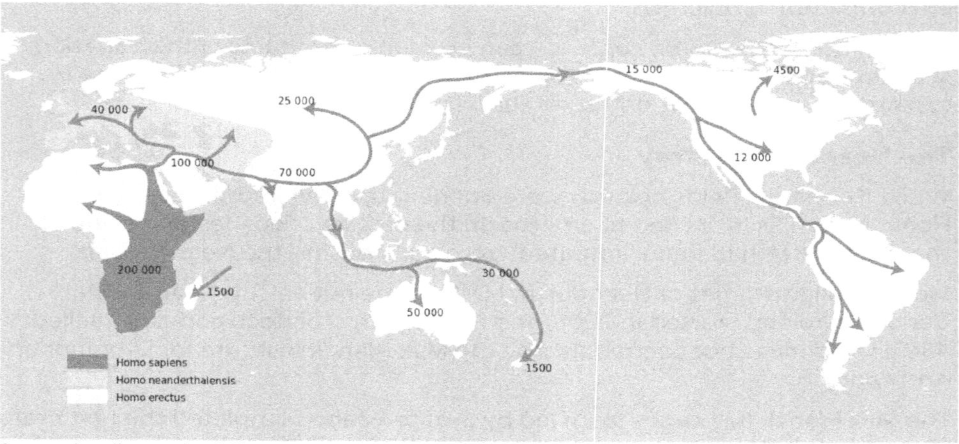

Successive dispersals of **Homo erectus** greatest extent (yellow), **Homo neanderthalensis** greatest extent (ochre) and **Homo sapiens** (red).

By NordNordWest - Spreading homo sapiens ru.svg by Urutseg which based on Spreading homo sapiens.jpg by Altaileopard, Public Domain,

Some scientists have proposed that superior human social skills and adaptability were the key the human success – the 'survival of the friendliest'.[42] You can't rule anything out as being possible.

And it could be true that larger Homo sapien social groups generated more new technology as well as more social skills, compared to smaller & more isolated groups of Neanderthals; so adding a technology advantage to a social advantage. Or maybe they could not adapt to climate change? But the scale of the disaster - complete extinction of Neanderthals – deserves a really persuasive explanation.

Given that whole branches of the family tree (including Neanderthals) have *vanished,* it was unlikely they were 'talked' out of the best locations and hunting areas, or persuaded by friendly discussions to leave. They must have been pushed out, involving force wherever friendly persuasion failed. Otherwise why would they leave?

There is no doubt that it is an area of genuine debate. But the most likely reason that our cousins *vanished* is due to the impact of Homo sapien 'competitive' activity. This is often just dropped into the list of many other possibilities, rather than being considered the most obvious starting point for research.

The best fitting theory remains that in the period from around 100,000 years ago (or even earlier) to around 25,000 years ago in Europe, Neanderthals were very likely to have been dislocated, forced out and eliminated from the whole European continent by Homo sapiens. This conclusion allows for some peaceful contact and interbreeding, but in the overall context of displacement.

For contemporary researchers like Rebecca Wragg Sykes, in her brilliantly researched 2020 book 'Kindred – Neanderthal Life, Love , Death and Art', the end of the Neanderthals was more complex, but still centred around competition for resources with Homo sapiens, which would have been intensive as the climate of the time got colder.[43]

And the same pattern seems to have been replicated in the Middle East and across Asia, with the vanishing of some much older travellers from Africa, including archaic hominins like Homo erectus, as well as Homo denisova, and even Homo floresiensis, a small archaic human from the island of Flores in Indonesia. Maybe it was the original *vanishing*?

Reinforcing this argument was a paper in the journal Nature Ecology & Evolution in April 2024, that also speculated that interspecific competition - competition for resources with other hominins, (or more generally individuals from other species) had played a greater role in extinction of our relatives than climate change.[44]

But a small part of who we are as Homo sapiens – 1% to 4% - is a legacy of that Neanderthal contact with humans.

DNA & Epigenetics

By the 1960's, following the discovery of the 'double helix' DNA structure in 1953 by James Watson and Francis Crick, researchers had found many portions of a genome sequence which had no discernible function. Or at least no function had been identified that produced an adaptive advantage. They called it 'junk' DNA, and it became a popular theory from 1960's. Use of the term 'non-coding' DNA was a more 'neutral' term for 'junk' DNA from the late 1970's, given the term 'junk' implied no functionality at all.

However, this perception of 'junk' was all changed by recent discoveries early in the 21st century, that raised fundamental questions about the theory of 'junk' DNA.

In 2003 the entire human genome was sequenced by an international consortium, and this gave us all the first 'cookbook encyclopedia' of where genes are located. And the search for what they do and what proteins they make began in earnest.

Then in 2012 along came the ENCODE project (a research program supported by the National Human Genome Research Institute). After 5 years research it reported that 76% of the human genome's non-coding DNA were studied and analysed – previously thought of as 'junk' DNA.

The Encode consortium had 442 researchers, at 32 institutes around the world, used 300 years of computer time, over 5 years to get their results & conclusions. In the process they examined 147 types of tissue, from cancer cell, stem cells, liver extracts and so on, and after a hundred different experiments, recorded which parts of the DNA code 'were activated in which cells at which times'.[45]

The ENCODE project uncovered many things, but a key finding was the importance of *regulatory* elements – the things that turn genes on and off.

Mike Snyder of Stanford University, the principal investigator in the Encode Consortium noted that there are more variants in the regulatory elements than in the genes themselves.[46]

The ENCODE project published 30 papers in leading science and biology magazines revealing researchers had mapped 4 million switches in what was thought, up to that time, to be 'junk' DNA. The concept of 'junk' DNA was dead.

In fact the DNA 'outside' protein-coding genes, the so-called 'junk' DNA, were actually a group of around 10,000 new 'genes' that code for components that control how the protein-coding genes work. The researchers found that around 18% of our DNA sequence is involved in regulating the around 2% of DNA that codes for proteins. Later researchers asked the obvious question – What was the other 98% doing?[47] This is a process involving complex interaction of hundreds of genes and their associated regulatory elements.

Earlier studies that compared the genetic code of people with certain diseases and the code of those without certain diseases (genome-wide association studies or GWAS) also largely matched up with the location of regulatory switches in Encode, so adding some independent confirmation of the research.

So here were the mechanisms that control or regulate genes - the regulatory elements. The liver cell knows it should make liver enzymes to metabolise alcohol, for instance. It knows that because one of these 'new' genes code for components that make this protein-coding gene in the liver works to make liver enzymes.

This is the mechanism of control that regulates this gene. Genes need these controllers or regulatory elements to work properly. If the regulators fail, then genes malfunction and this could result in diseases like diabetes, Crohn's disease or cancer.

Out of these breakthroughs and subsequent work came the term *Epigenetics* – the study of heritable *phenotype* (eg. a characteristic like blond hair) changes that do not involve alterations in the DNA sequence. The Greek prefix (epi) means over or outside or around. In epigenetics it means features in addition to, or on top of, the traditional genetic basis for inheritance.[48] It mostly relates to changes that affect gene activity and expression.

In the end, it seems that big parts of what was previously called 'junk' DNA are involved in this 'epigenetic' regulation.

So the 'set of instructions' in DNA are not the whole story. It's more like a 'script' in a play. Nessa Carey, a prominent British biologist working in the field of molecular biology and biotechnology, wrote the 'The Epigenetics Revolution', and in this she compared the role of epigenetics, to thinking of the Romeo and Juliet play by Shakespeare.[49] The 1936 'Romeo and Juliet' movie by George Cukor was entirely different to the 1996 'Romeo and Juliet' movie by Baz Luhrmann. Same starting point, but very different.

The DNA didn't change. Yet things go differently in a different environment. An example she quotes is of abused or neglected children ending up with a much higher risk of adult mental problems than children without that experience. It is something that the much underrated 'common sense' would tell us is so. And here at last is a framework to support it – epigenetics.

Nessa Carey describes this kind of case as an example of *epigenetics* – a new discipline that is revolutionising biology.

She defines epigenetics as the set of modifications to our genetic material that change the way genes are switched on or off, but don't alter the genes themselves.[50]

This is the molecular description of epigenetics, which we are focussing on, not the more general use of the term. 'Epi' comes from the Greek meaning 'at, on, to, upon over or beside'. Regions of our DNA can be 'switched' on or off by molecular amendments that do not change the underlying genetic code. Chemical groups are added to DNA or associated proteins, or removed, and this changes the 'expression' of nearby genes (ie turns them on or off).

It now seems that epigenetics was a 'missing link' in understanding biology in general and evolution in particular. Large areas of biology are affected by epigenetic mechanisms, and Nessa Carey described is as a revolution in our thinking.

In fact, she describes epigenetics as likely to become almost a scientific 'signature' of the 21st century.[51] Darwin & Mendel defined the 19th century as the era of evolution and genetics; while Watson & Crick defined the 20th century as the era of DNA, and how genetics and evolution interact; and the 21st century will be the era of the new scientific discipline of epigenetics.[52]

Epigenetics as Code

Of course, when reading about the emerging importance of epigenetics, many programmers will note the similarity with much of the coding they do.

With computer software, there is the program, then there are things that control the program. It is a fundamental way software works. Otherwise, you would need to have different software to do different things and a lot more complex coding.

So how is this problem solved in computer software development? Simple. You have one program (say for example a piece of accounting software). But can the *same* software work for a sole trader - say a cake shop – as well as a multi-national corporation – say a Bank?

If the CORE software functionality was strong, then in setting up the system for the cake shop owner, the implementer would use what we call 'site settings' or 'parameter settings', while installing and setting up the system.

He/she would set the multi-currency and multi-lingual, advanced journals, complex screens and advanced reporting to 'No'. With the right design, even certain fields on one screen can be turned 'off', and disappear, with the change of a parameter or site setting.

Then in *theory*, once these functions are turned 'off', what would appear, is a simple system for a cake shop owner.

Now in *practice* the Bank would probably get a much bigger system, as they would want to start with a set of advanced functionality; which they would never want to turn off, so a different system (to the cake shop) would make sense.

But the principle still stands – the process being described as using 'site settings' or SWITCHES is exactly how DNA functions.

A site setting is a more 'global' setting – like language = English, for example, or default currency = '$US'. It is a 'big picture' setup item, best done on installation and then left like that. (Note that another 'division' of the same database could still be set as language= Spanish, for example, and default currency = 'Euro').

But inside the software are other settings that might be called 'parameters', that also allow for change. For example, in the accounting system, you can define the accounting year as going from July to June the following year, a financial year, or an ordinary calendar year. Or would you like dates to appear as MM/YY or YY/MM? There are usually a large number of these parameters, adding great flexibility and adaptability to the software.

New Fields of Computing in the Biology area

Interestingly, in a case of convergence with the world of biological software, there is already a development method called 'evolutionary computing', in the computer software world.

Evolutionary computing is software to optimise change by looking at patterns of biological evolution – essentially using machine code to mimic the way evolutionary changes come out of biological code. It emerged from pioneering work of brilliant researchers in the 1950's, 1960's and 1970's including Fogel who first talked about evolutionary programming[53], and others like Alex Fraser & John Holland.

The concept in evolutionary computation is that 'candidate' solutions are generated and iteratively (repetitively) updated. By this method less desired solutions are removed, and small random changes are possible.

It is based on Darwinian evolutionary theory, where outcomes change due to natural selection and mutation. Coded outcomes 'evolve' and becomes 'fitter' – producing the most efficient (optimised) solution – with real world & difficult engineering problems like designing a robot or finding the best way to mine data. German researcher Ingo Rechenberg, also one of the pioneers of the field in the 1970's[54], applied evolutionary computing to aerodynamic wing design.

More use of evolutionary algorithms (as part of evolutionary computing) could make some businesses more successful in the commercial world by building more 'intelligence' into their approach to solving problems.

Other related fields of study are now also emerging fast – computational biology (computer simulations used to see how biological systems and their data work), and bioinformatics (using software to collect and analyze biological data), currently seen as sub-branches of biology (that can also be regarded as part of the software world).

By seeing many things as code changes means we are free to find explanations for changes outside the current paradigms and cultural or intellectual frameworks. We can seek an explanation based on recognising that we are talking about - at the fundamental level - is code.

A/ EPIGENETICS AND HEALTH

Seeing DNA as code is so logical. There is really no other way to describe it.

But it's not ordinary computer code. It can be described as biological software code. Code of a simplicity and limited tool set, yet with complexity of outcomes we still are a long way from understanding.We have discussed the Darwinist 'natural selection' process of *selection, variation and inheritance*. And we have also reviewed the epigenetic revolution, the so-called 'switches' in the (what used to be called) non-coding DNA area that activate or silence the expression of genes.

The software design of DNA with its epigenetic switches is a logical product of evolution; it adds a lot of adaptability that living things need to survive, and makes that survival become more possible.

Epigenetic effects can now be considered as an integral part of the way humans respond to the impacts of the external world – whether the physical or social environment.

Important to this are the possible epigenetic effects of an external environment 'clicking in' or 'switching on' with long term and inter-generational effects.

The most obvious & visible impacts of epigenetic effects is likely to sit in the massive area of human health, based on research done so far.

We are just at the beginning of understanding epigenetic coding and its impact on DNA and human behaviour.

The story of Audrey Hepburn (from the movie 'Breakfast at Tiffany's) is probably the earliest proven and well documented example we have of this today.

Audrey was a survivor of a WWII event called the Dutch Hunger Winter. At 16, from 1944 to early 1945, she endured starvation and poor health as the war grinded to a finish. 20,000 died until food supplies were restored in May 1945. The Dutch survivors were a well-defined group that epidemiologists followed to study the long-term effects of the famine.

The babies born small, & stayed small all their lives, with low obesity rates. Their bodies never got over the early period of malnutrition. Children of mothers who were malnourished early in pregnancy, had higher obesity rates. Then the grandchildren of the mothers seem to follow the same pattern – effects were passed on to future generations.

The implications of this well-documented case study are epic. That environmental conditions can lead to non-DNA coding turning 'on' or 'off' switches that change the way DNA works in human beings, and that it can be inter-generational.

In addition to this case study, great progress was made by 20th century researchers with many disease-causing genes in DNA, causing abnormalities that are passed down the family tree. This includes well known conditions like Cystic fibrosis, Down Syndrome, Huntington's disease, and Haemophilia, as well as lesser-known conditions like Sickle cell anaemia and blood conditions like Thalassemia.

Chronic diseases that affect humans were the first place biologists started to look, when considering the impact of epigenetics.

A big issue today is overeating and obesity; a feature of 'modern times', along with the rise of so-called 'junk food'. Or if you like, ironically 'junk food' being affected by 'junk DNA', to use their common descriptions. What epigenetic impacts of this rise of junk food lay ahead? What health impacts will come from a generation of 'fat kids', as they are sometimes called. Will a generation of young kids one day start exhibiting health impacts from epigenetic 'switches' turned on or off as they grow up on a diet of coke and hamburgers?

Then there is the case of folic acid, known as a supplement for pregnant woman and in reducing the occurrence of spina bifida in babies. Folic acid is being explored for more epigenetic effects by researchers. It has already been shown to be powerful and effective. What else can it do?

Cancer researchers are investigating the use of histone deacetylase inhibitors (HDACi) in our diets, with the view of preventing cancer.[55] HDACi can act by altering gene expression to help stop the growth of cancer cells.

Epigenetic enzymes may have more complex roles in influencing cells than we know right now, and Nessa Carey speculates that some enzymes originally had other functions, but have been 'press-ganged' by evolution into becoming controllers of gene expression.[56]

All this points to the Health area being a hot topic for new research in the coming decades.

Epigenetic 'switches' have a role in causing various illnesses, triggering various conditions. This century we will learn the extent to which epigenetic switching affects our health. And beyond impacts on health, a whole new area is opening up.

B/ EPIGENETICS AND HUMAN AGGRESSION

As discussed, the impact of epigenetics on human health is now established,

What is the next frontier of epigenetics for humans? There is an obvious candidate for study, a characteristic of human behaviour that has been troubling many for generations – human *aggression*.

Aggression can be defined as behaviour intended to cause physical or mental harm. Researcher & author Richard Wrangham breaks down aggression into 'proactive' and 'reactive' aggression. Something like the difference between premeditated, and impulsive aggression. He believed that in humans, proactive aggression stayed high, but reactive aggression was suppressed by 'self-domestication'.[57]

He saw self-domestication as something we share with domesticated animals, where we have built up social tolerance and low emotional reactivity to provocation. This developed as advanced human use of *language* allowed genetic selection against alpha males' propensity for reactive aggression. Basically, subordinate men ganged up on male bullies and plotted their death behind their backs.[58] It's a fascinating theory.

Domestication is seen more like a permanent genetic modification than an epigenetic change, and could explain the fall in 'reactive' aggression. But the origin of 'proactive' aggression, and its important ongoing role in human society is still a mystery.

A theory about human aggression is that constantly unpredictable climate in Africa helped create a drastic urge to survive and a willingness to move to do so, and take by force new territory, based on expanding weapons technology. An ape ready to invent new weapons and prepared to kill anybody in their way? It's a harsh view.

The argument could be that environment pressures led to some epigenetic 'switches' to turn off, so that compassion became qualified, and other switches to turn on, so that 'proactive' violence & aggression, especially for men, became acceptable in most forms – for the tribe, the race, for beliefs, for god, and for the family – and maybe most of those switches remain that way today.

It is the start of understanding some human behaviour to propose that it has epigenetic roots, where somehow feelings of ethnic or religious or ideological superiority justify killings and aiming to exterminate other groups. An epigenetic explanation suggests that genetic switches get turned on in some circumstances and then suddenly are temporarily gone.

We are just at the beginning of understanding epigenetic coding and its impact on DNA and human behaviour.

If epigenetic changes resulted in changes in the aggression levels of Homo sapiens, it could explain a lot.

It does not rule out explanations of human aggression that involve *sociobiology* (explaining social behaviour in terms of evolution)[59] or *evolutionary psychology* (seeing human behaviour from an evolutionary perspective, looking for *psychological adaptions* that evolved over time to solve any physical, social, economic or environmental problems).[60]

It is early days, but this area of study (human aggression) does need to be given more priority by universities and governments, given its importance in shaping the lives of millions.

The Genocide Story

Jumping forward from the assisted *'dying out'* of our distant relatives, we have more recent examples of human behaviour that is also very disturbing.

Many think that *genocide* is a rarity in human history. If only it was. There is a case that it is one of the defining characteristics of Homo sapiens. Sheer frequency of these events is shocking but is rarely paraded as Homo sapiens *signature or most recognisable* behaviour.

Author Norman Naimark in his landmark book 'Genocide A World History', makes the point that *genocide* has been part of human history from its very beginnings.[61]

Genocide is defined by him as the destruction of a nation or of an ethnic group, from the Greek *genos* (race, tribe) and the Latin *cide* (killing). But the term can be used more loosely to define the murder of whole communities with different religions or political beliefs, if we call these groups *'tribes'* of a kind. And the term can be even stretched to cover murder of other human species, given the absence of another word to describe this. Interestingly, to be genocide, the act does need to have the *intention* to destroy part, or all, of a particular group....it is not just the outcome that is important.

Norman Naimark makes the point that lack of evidence in pre-modern times really means it is difficult to come to 'hard-and-fast' conclusions[62] about events the further you go back in time.

So the probable genocidal extermination of the Neanderthals (and more ancient relatives in Africa) could never be proved, and becomes a conclusion to be made based on the circumstantial evidence, that it is the most likely and most logical explanation. Essentially, Homo sapiens were at the scene of the crime and all other explanations are less plausible.

More detailed written evidence of genocide starts from the ancient world starts from the destruction of Troy around 1200 bce [63](before common era), and then follows Homo sapiens through the ancient empires, to the role of European conquerors then settlers in the mass murder and genocide of hunter-gatherers, to the modern genocides like the Armenian[64] mass murders, (still denied by some), to the Nazi Holocaust of the 1940's[65], (still denied by some), the Communist genocides (Stalin, Mao), then more recent genocides like the Rwandan genocide in 1994 and the tragedy of Kosovo (1998-99).

The other author who researched genocide in depth was Jared Diamond. He became a professor of Physiology at the University of California Medical School, and is well known for his contribution to the fields of ecology and evolutionary biology.

His book 'The Rise and Fall of the Third Chimpanzee' (1991) was influential in placing human beings as gifted creatures; but he also reminded us of the darkest parts of our animal origins.

Part Four of this book (Chapters 13 – 16) is an in-depth consideration of, as he calls it, 'our propensity for killing of other human groups'.

Xenophobia is the term he uses to describe this behaviour - extreme fear and dislike of foreigners, or people from different cultures, or strangers.

This can lead to genocide, or attempted genocide, or simply mass killings that don't have a rational cause.

In his Chapter 16, he really explored genocide in all its shocking detail. It seems that our always improving technology led to better and better weapons, (in particular, the ability to kill at a distance). And in many cases, *xenophobia* seemed to end up with murder – mass murder.

So he ends up with tables of 'genocides', covering mass killings from the earliest times up to today.

They include 8 case studies since 1492 where more than one million were murdered each time. These include 8 cases of more than a million murdered – the Nazis killing Jews, Gypsies, Poles and Russians in 1939-45, the Russians killing political opponents 1929-39, the Turks killing Armenians in 1915, the Americans killing American Indians 1620-1890, the Spaniards killing Central and South American Indians 1498-1824, the Spaniards killing Caribbean Indians 1492-1600, the Pakistan Army killing Bengalis in 1971, and finally the Kymer Rouge killing Cambodians 1975-79.[66]

He also found another 8 cases where more than 100,000 were murdered, which may go up to much larger figures, that can't be verified, including Croats killing Serbs 1941-45, Russians killing ethnic minorities 1943-46, Hindus & Moslems killing Hindus & Moslems 1947, Australians killing Aborigines 1788-1928, North Sudanese killing South Sudanese 1955-72, Idi Amin killing Ugandans 1971-79, Tutsi killing Hutu 1972-1973 and Indonesians killing communists and Chinese 1965-67. Some of these cases may also exceed a million people murdered.[67]

For Homo sapiens, mass murder and genocide are not a rarity. Is this level of aggression now hardwired into us by epigenetic switching in response to major events in the history of Homo sapiens, or is it caused by events in society or the environment that do not have a genetic explanation?

Whichever the case, the whole area of human aggression and how it manifests itself desperately needs more research to know whether we will always be like this.

C/ EPIGENETICS AND ENVIRONMENTAL DAMAGE

If epigenetic switches help explain human violence and aggression, then can the same explanation be offered to explain environmental degradation and the extinction of other species?

All other living creatures on the planet have been affected by the rise of Homo sapiens. Around 83% of wild mammals have been lost in the last 50,000 years[68]. Those figures don't take into account human impact on the oceans or on habitable land for other creatures. Human domination of the planet has been done at the expense of *'replacing'* many species and previous ecosystems.

The 2023 IPCC (Intergovernmental Panel on Climate Change) report on climate change found that impacts on people and ecosystems are more widespread and severe than expected, and future risks will escalate rapidly with every fraction of a degree of warming.[69] Earth's temperature could end up showing a three degree increase above pre-industrial levels as soon as 2099.

It's a crisis of global warming that is already affecting millions of people.

Author Adam Kirsch's observation is that humanity's domination of the planet is so extensive that evolution itself should be redefined. In his book 'The Revolt Against Humanity', he offers a new understanding of the term *'survival of the fittest'*. He thinks that the term should now be defined as *'survival of what is most useful to human beings'*.[70]

Antonia Guterres, Secretary general of the United Nations in 2020, described it as humanity waging war on nature, with biodiversity collapsing and a million species at risk of extinction as ecosystems disappear.[71]

The author who really put this brilliantly was James Lovelock, in his controversial 2009 book *'The Vanishing World of Gaia'*, who said we are what our genes made us, as an explanation for the massive destruction of wildlife and forests.[72] The concept of a vanishing environment is powerful – and alarming.

Human aggression and environmental destruction can be the starting point for epigenetic research into human behaviour.

D/ THE MICROBIOME AS A FACTOR

If incorporating epigenetics into any theory about human evolution is hard, consider the new territory of the *microbiome*.

The microbiome are microbes that colonised us, including bacteria, viruses and fungi. They can affect our feelings, orchestrate the immune system, and influence how the brain works and grows.[73]

These are the trillions of other organisms that live in the human body, and form part of our daily life. Some scientists argue that each individual should be considered a *'metaorganism'*, due to our extensive interactions with microorganisms. This is now an area of major research, with its own collaborative research centre.[74]

Up to now most scientists see 3 discrete parts of our *self* – the immune system, the genome, and the brain. But they are all affected by microbial impacts, according to Thomas Bosch at Keil University in Germany[75]. Very little of which we understand, as far as how they change our minds, or mood or our health.

Is the impact of the microbiome similar to that of epigenetic impacts, where each individuals interactions with microorganisms mean we need to rethink what it is to be human, and what is influencing us? How have humans and their microbiomes co-evolved, and what is the impact on human behaviour. Do they affect fitness in survival terms?[76]

Are they acting like switches that turn on and off behaviour and even genes themselves, so they end up with inter-generational impact? Especially given that it is argued by some that 30% of genes in the human genome have a bacterial origin.

The role of the microbiome sounds like it has coding implications for human evolution. We need greater understanding of how the microbiome has changed over time, but that may not be easy to find.

It is early days, awaiting research of the whole area, but maybe this area is as important as epigenetics?

E/ DIVERGENCE AS A FACTOR

An additional explanation of some human behaviour comes from Tony Ord.

Tony Ord is a philosopher from Oxford University, who wrote the ground-breaking work 'Precipice' in 2020. He defined the problem as the rapid growth of humanity's power outstripping the slow and unsteady growth of our wisdom, creating a divergence.[77]

He felt that slowing technological progress would give us some breathing space, and allow our wisdom to catch up, to limit this divergence.[78]

And it is hard to see any signs of change, any sign that the divergence is vanishing.

In fact, there is a case for saying that the divergence continues to expand between our slow growth in wisdom and our fast growth in technological knowledge and power.

F/ 'CODE' IS NOT JUST A WORD

Charles Darwin talked about *'survival of the fittest'*. The phrase was first used by Herbert Spencer in 1864 and was adopted by Darwin in 1866, as a useful phrase to explain some ideas from his book 'On the Origin of Species by Means of Natural Selection', that came out in 1859[79]. Alfred Wallace (Darwin's contemporary and independent co-founder of the overall evolution theory), suggested Darwin use this phrase as he felt the term 'natural selection' was too 'indirect'.

The *'survival of the fittest'* phrase is a much-quoted rough summary of Darwin's thesis, but leaves out a lot of more subtle and clearer analysis about how 'natural selection' works. The *'survival of the fittest'* phrase is disliked by many for its over-simplification and confusion the expression can create. As Richard Dawkins noted in 'The Extended Phenotype'[80]; 'fitness' is a very difficult concept, and can have many meanings. But the phrase somehow stuck in the public mind.

Of course, the fact that the book 'On the Origin of Species by Means of Natural Selection' was released in 1859, meant it was published long before the emergence of modern computers into popular culture, and awareness of the power of programming and the role of software. It was equally so also when James Watson and Francis Crick discovered the 'double helix' or 'twisted ladder' structure of the DNA molecule in 1953.

If Darwin wrote the book today, what he might be discussing would be the 'survival of the fittest ***code'***. Is that a very different equation? Yes, because the focus is about code. And as we have found; discussing and understanding 'code' is critical, as DNA is based around code.

Richard Dawkins original work 'The Selfish Gene' came out in 1975, and has been a milestone in understanding evolution, in particular the evolution of the gene, as well as the whole organism.

Yet in the 30th anniversary edition, even he was wondering about the title of his book, in particular the word 'selfish'. His publishers told him he could touch little in the new edition, but his friend (publisher Tom Maschler) wondered if he should have called the book 'The Immortal Gene', to covey the central idea about the immortality of genetic information.[81] He concedes his friend may have been right, which shows an openness and honesty to his credit. Tinkering with the title does not alter the content of this milestone work.

Yes, by 1976, business & scientific *computers* had arrived, and had already helped take Neil Armstrong to the moon in July 1969. But software was not seen as being at the heart of what was going on around us. It was still 'emerging' as something that would start to reshape our thinking and force a rethink of our past.

But if it had been re-thought; Richard Dawkins might have not called his book 'The Selfish Gene; but 'The Selfish **Code**', or 'The Programable Gene', or 'The Immortal **Code**'. Just as fascinating, and starting to tell another story behind the story.

'Code' is not just a word; it is a word with the power to change the nature of the whole discussion. It is *the* word that is the key to explaining much of what is actually happening.

G/ SUMMARY

The original pioneering work of Darwin and those who followed led to a new understanding about evolution.

The discovery of the role of epigenetics is helping explain human behaviour and is starting to fill gaps in Darwin's original theory.

'Junk' DNA, DNA that made no sense until epigenetic regulation was added into the equation, is now beginning to be understood.

It is a new perspective on the importance of code in evolution - that it is all about *biological software.*

But there is still a lot we don't understand about human behaviour, such as the reasons for aggression, and the causes of environmental destruction. The epigenetic revolution could help here. Social & psychological insights are also likely to be part of explanation.

But humans are complicated. Maybe we are stuck with being Leonardo da Vinci, Mother Theresa, Bill Gates, Marie Curie, but also Joseph Stalin and Genghis Khan all jumbled up in a crazy mix?

It's insulting and terrifying while also intriguing. It's like looking in the mirror and seeing someone else. It's an *'out of body'* experience, to suddenly realise that we aren't *exactly* the scholars, the compassionate and intelligent creatures we think we are.

We can also kill, we can abuse, we can destroy, but mostly we replace; some of those things on a continental scale. We are divided by artificial tribal, cultural national, religious, racial or ideological constructs in our heads that often lead to violence and death. *Replacement* is one of hallmarks of Homo sapiens.

We need to face the likelihood that we, as a species, are *aggressive* more than *wise,* while at the same time suffering from a major problem of *divergence*, where fast moving technology and military capability is easily surpassing slowly growing human wisdom. And the gap seems to be getting wider as technology is still moving very fast.

We need to let go of the stereotype of Homo *sapien* – that we are this *wise human.*

That wise creature never actually existed in reality, was always less intelligent, yet at the same time more complicated, aggressive, and destructive than at first sight.

And now after rethinking evolution, this wise creature is destined to *vanish* in our imaginations, just like it did in real life.

4. Rethinking the Brain

The way the brain works has always been a mystery.

In the past, it has always been defined in terms of the technology of the time – working like a clock, or a machine in a factory. Which never really made much sense.

Then at the dawn of the computer age, the brain was often seen as working 'like a computer'.

The Brain is Like a Computer?

Since the emergence of modern computers during World War II, then especially in the 1950's, 1960's,1970's and 1980's; an idea of popular culture was that the *'brain is like a computer'.*

When considering *'computer related'* brain functions, there several key brain functions that are an obvious *'analogy'* to how a 'computer' works:

It can collect information via eyes, ears, touch, smell etc (input)

It can evaluate new information (processing)

It performs many tasks, like making decisions, calculating, planning (output)

It can store & report on information (short & long-term memory)

All of these functions are common to many living things. Input and output are something all living creatures do. The complicated part of the process is the 'processing'. The equally fascinating part is the role of memory, storage that allows memories and the accumulation of learned experience.

But as early as 1951, neuroscientists like Karl Lashley argued against using an 'machine-based' metaphor.

Then as time went on, the idea of the 'brain as a computer' started to really annoy more researchers in the neuroscience world, and the whole idea became unpopular in some quarters, as being too 'simplistic'.

There was a growing recognition of the brain's complexity. For instance, Gerald M Edelman & Giulio Tononi in their excellent book 'Consciousness' describe the brain as being one of the most complicated objects in the universe and one of the most remarkable structures to have emerged during evolution.[82] They note that the adult human brain weighs about 3 pounds and contains about 100 billion nerve cells, (neurons).[83] But they also make the clear and sharp comment that the brain is NOT a computer.[84]

You can see that this topic generates some strong reactions!! The main argument those researchers put is that the brain has a structure and organisation that means its functioning does not seem to fit with the idea that it follows a precise set of instructions or performs computations.[85] This is because synaptic connections keep changing and there is massive individual variability in how each brain works.

So, the original idea that *'the brain is like a computer'* analogy has been struggling, hovering like an uninvited guest who turns up at a birthday party; a victim of the simplicity of the argument and also the failure of writers to clearly distinguish between 'hardware' and 'software' and its many forms, (and also 'operating systems', the way software 'talks' to the hardware).

Still, as an analogy, it offers a clue.

The Mind is the Software of the Brain?

A better formulation of this topic came from cognitive scientists of the 1980's and 1990's, as they grappled with how to explain the *mind.*

Instead of saying *'the brain is like a computer'*, some were saying *'the mind is the software of the brain'.*[86]

'Philosophy of mind' researcher Ned Block felt that the *computer model of the mind* was profoundly un-biological, in that the computer model of the mind abstracts away from the biological structures in the brain.

At the time Ned Block wrote this (1995), the *biological* world and the *computational* world must have seemed far apart, and his observations reflect the chasm seen between the 2 worlds at the time.

The importance of this study in 1995 was that it showed how leading thinkers at the time like him were starting to explore the general idea that *software* is part of the way the human mind works – and also starting to wonder how a 'machine model of the mind' would work.

In his article Ned Block argues presciently that looking at the biological and computational approaches would lead to *discovering* the program of the brain. He was a pioneer pointing us in the right direction at a time when the full weight of the DNA revolution was just starting to roll over society, and the internet was emerging as a force in computing, and modern software was moving ahead very fast, branching into many areas unheard of in the 1990's.

Ned Block went on from the paper in 1995 to become a leading academic working in the philosophy of the mind area, making big contributions to our understanding of consciousness, (distinguishing between 'phenomenal' and 'access' consciousness), and continuing to find weaknesses in the Turing Test (the Blockhead argument). Such was his influence that in 2019, a book of essays on his Philosophy of Mind and Consciousness, called 'Blockheads!'[87], was released, a tribute to his ongoing but lasting impact.

From the 1990's onwards the focus moved onto researchers focussing on the development of artificial intelligence (AI) software (by developing software algorithms to become intelligent), or alternatively, by development of neurological networks, (that try to copy the brain's structure in software).

Yet despite decades and decades of research since Ned Block's article, our understanding of the brain and things like 'intelligence' & 'consciousness' remains a *frontier* of brain study. This is despite the fact that more and more information from thousands of researchers is becoming available about the structure & functions of the brains of many creatures, including humans. We also know a lot more about what parts of the brain do *which* functions.

Yet even as this data accumulated, many in the area were still searching for a 'theory' that explained the brain. German neuroscientist Olaf Sporns put this brilliantly in 2020. He noted that the science of the brain (neuroscience) still largely lacks *a theoretical framework* for converting brain data into fundamental knowledge and understanding.[88]

British neuroscientist Mathew Cobb also observed that some neuroscientists could not see a clear path ahead to understanding how the brain works, and it was hard to see where they should be going.[89]

That is, despite more and more information and facts emerging, it is still a struggle to understand the brain.

Leading thinkers in the area are searching for an alternative way of theorising about the brain, leaving the path open for new ideas.

The Brain is Software?

So, it is of no value to return to the old idea of seeing 'the brain is like a computer'. It can now be seen as just a simple analogy, a potential guide.

And while the idea that 'the mind is the software of the brain' has some potential, in that it introduces the idea that software has a role in the way the mind works, but it still leaves us thinking about the brain as a complex black box without a theoretical framework explaining it.

However, an alternative explanation about how the brain works is available – that the brain is a piece of software: *biological* software.

As outlined in Chapter 2 – Rethinking DNA, the operations of human cells are in fact better understood as software of a different kind – biological.

If *cells* are part of biological software, using the DNA code for creation and evolutionary changes, then how does the brain work? What kind of software is it?

In particular, how does it lead to the most complex area of human behaviour; the higher brain activity like reasoning and planning?

If this theory proposed here is right – then we are looking at how *advanced software* functions in this area operate.

It is going to be advanced software that nobody fully understands now.
For instance we are still looking for good explanations of things like 'consciousness'.

Our purpose here is to reset the starting point for discussion about the brain. It is proposed here that it is operating as biological software. But how?

How does the brain work?

As noted above, the way the brain works is one of the world's enduring mysteries, and (most think that) nobody has proved an overall theory of what happens yet.

But it is such a challenge to unravel its mysteries that there are always new ideas and projects around to solve the riddle. Europe's Human Brain Project[90] and the International Brain Initiative (IBI)[91] are serious efforts led by neuroscientists to understand how it works. These are amazing projects in motion, and it will be fascinating to watch their progress.

Of special interest here is the Human Brain Project's work in neuromorphic computing. This implements aspects of biological neural networks as copies on electronic circuits, aiming to understand the process of learning and development in the brain, and applying 'brain inspiration to generic cognitive computing'.[92]

In short, that Project aims to look at the brain and mimic the biological networks as a way to improve the way generic 'cognitive' (based on studying the mind) computing works – the neurological network approach.

This 'neuromorphic' computing approach to explain how the brain works – essentially mimicking the brain - may produce a breakthrough. It hasn't happened yet.

The other main approach to understanding the brain is the approach based on artificial intelligence – building algorithms that can 'learn' to think. Also a massive project.

However, as mentioned before, there is plenty of scope for new theories, as the brain still holds its secrets, and there is still doubt about where intelligence and consciousness come from. There is much that can be rethought & reimagined about what we know and don't know.

Figuring out the circuitry inside the *neocortex* (the large part of the brain involved in higher brain functions linked to intelligence, like language, planning, reasoning, cognition, vision, hearing, touch), is at the heart of understanding what goes on.

Jeff Hawkins' 'Thousand Brains Theory' (2021) is one such fascinating theory about how the brain works. His theory is that as the brain learns models of the world, these models are built in the neocortex area of the brain using hundreds of thousands of map-like 'reference frames', which are manipulated by the brain to plan and think.[93] Thus the name 'Thousand Brains Theory'. The fact that there are thousands of models and thousands of individual thoughts each moment, means the brain somehow conducts a vote on these models to decide what to do. Your perception is the consensus reached by voting.[94]

The 'Thousand Brains Theory' is a real insight into human decision making, and also an explanation of how prediction works based on brain models. It doesn't give us all the answers, as the way *decisions* are made still has a lot of questions about it.

Another new idea about the brain is *emergence*. Some say consciousness can only be explained by *emergence*, rather than only looking at its constituent parts.[95] So in this explanation, it is like the *wetness* of water, that isn't an obvious outcome of all the properties of water when looked at one by one; but is an emergent quality.

Or maybe consciousness just *flows* rather than emerges. In the work called 'The River of Consciousness' (2017), the last book by Oliver Sacks, the 'river' is 'time', which is continuous, rather than a succession of discrete moments, like beads on a string.[96] Use of *river* analogy may partly come originally from William James, in his 1890 'Principles of Psychology', where he introduced the term '*stream* of consciousness'. That use of the word 'stream' is there to also describe a continuous flow of thoughts, like a river – a flow of consciousness.[97] We will come back and look at consciousness in more detail in Chapter 5.

Yet another creative idea about how the brain works was put forward in 2023 by well-known cognitive scientist Andy Clark, in his book 'The Experience Machine – How our minds predict and shape reality'.[98] The book is really about the brain's ability to make and change predictions about what we expect to see in our daily lives. His idea is that the brain is basically a giant prediction machine.[99] It is a strong presentation of this view, and persuasive.

Jeff Hawkins theory that the brain builds 'models of the world' - is similar in some ways, in that he sees *prediction* based on these brain models. As Andy Clark describes it, the brain learns and maintains a model of the body and world, and uses it, moment by moment to predict things, whether it is the absence of a birdsong in the morning[100], or where to walk in the dark.

The impressive thing about Jeff Hawkins work is that he tries to tie his theory back to the actual behaviour of neurons (he discovered that predictions occur inside neurons in 2010)[101], while looking at software simulations to test his model of neuron behaviour.[102]

But these theories still don't define the activities of the brain as 'biological software'. This biological software approach is using a different, lower-level starting point for analysis, based on going back to the fact that all that happens in the brain is based on code-based biology; whereas much of the focus of the work of Clarke & Hawkins is to try to come up with a theory of intelligence. Accepting the brain as biological software would mean that Andy Clark's book would not be called 'The Experience Machine', it would be called the 'The Predictive Software of the Brain'.

That's why the implications of seeing the brain as *biological* software are profound. So where is our conscience? Where is our soul? Where is our compassion, love, hate, depression, excitement? And where is our sense of right and wrong?

If they are biological software utilities influenced by external events, does that *diminish* them in some way, compared to how these things were imagined in the past? Are they reduced from being a higher part of our nature and brain, to being *software* open to change, with no moral compass or innate sense of right or wrong playing a role in our story? Hard questions with no easy answers, once we consider the whole brain as software well beyond our current level of comprehension.

We are still a long way from fully understanding how the brain works, and how decisions are made.

If figuring out the science of 'why' and 'how' we make decisions is complex and not fully understood, it is also telling us that the process of the decision making inside our brains is not straightforward, especially if *prediction* is a core decision-making tool.

It goes some way towards understanding the modern-day mystery of why all of us occasionally make bad decisions in our lives.

The Power of Memory

Another important part of the brain's decision-making ability is the storage of information – memory.

Memory is a vital part of brain operations. It can be split into short term and long-term memory.

Some things we remember for a short time, like what we had for dinner a month ago, or what time we got home last Wednesday. Other things we remember for ever, like when our mother or father died, or when a war started, or we lost our job. They are all experiences stored in the brain.

Remembering things – like how to build a tool or weapon, how to cook a meal, make some clothes, or where to find game or plants to eat – is an essential skill for any living thing. Human memories can be communicated by language, and so knowledge is spread and passed down generations, and not lost.

But being able to *store* memories and facts *outside* of the brain, has been a central part of the human success story – effectively 'speeding up' the remembering process, by recording it and making it independent of any individual's memory.

The *permanence* of writing is the significant factor here. It means less re-learning, and faster dissemination of new and existing knowledge. The development of *writing* put language into solid, permanent form, meaning *knowledge* could be passed on, and did not have to be re-discovered.

This combination of language and writing and storage has put the human race on a rapid learning curve compared to all other creatures.

The world's oldest writing was discovered in the 1990's in an Egyptian desert cliff, even older than writing found in Sumer (now Iraq), and has been dated at around 5,250 years ago. The inscriptions and symbols apparently tell the story of a King's victory, the world's first document. John Coleman Darnell and his wife Deborah, both experts on ancient Egypt, made the discovery of the tableau in 1995, at Gebel Tjauti, about 400 kilometers south of Cairo.[103]

Challenges to this claim keep coming up. In 2023 experts researching European cave markings from around 10,000 years ago, found codes that hinted at the early origins of writing too. Drawings of bison, deer and horses featured graphic symbols like lines, crosses, asterisks, and dots that could relate to various animals breeding seasons, so allowing this information to be passed on the many generations.[104]

The last 5,000 years has seen the evolution of reading and writing, from being a preserve of the wealthy, to become something for us all, a fundamental turning point in learning, and dissemination of knowledge.

In 1250, Korean Civil Minister Choe Yun-ui printed the first book using metal blocks placed into a frame for each page and then inked. The fact Korea at the time was under invasion, and the Korean language featured so many characters, slowed down the impact of this invention of printing.[105]

Then in 1440, Johannes Gutenberg re-invented modern printing. He also combined moveable metal type with a press, to open up the production line for books. Books for everybody, eventually.

Writing has helped drive new inventions, improvements based on past knowledge, and development of higher culture – sometimes at least.

Then followed the Internet, which has also taken books to a whole new environment, and changed the way they look, are reproduced, and how many can see/read them. It has facilitated the spread of knowledge like nothing before it.

The combination of human ability to speak, has turned into writing, and become *stored* knowledge that has accelerated the accumulation of knowledge and human domination of the planet.

The role of technology in transforming and extending biological human memory into physical and computer storage is one of the most amazing and under-reported chapters in the spread of humans across the planet.

Conclusion

How the brain works is a *complex* area, with a lot more discoveries to come, with a lot of discoveries needed before we truly understand it.

In Chapter 5, we look at things like intelligence, sentience & consciousness, to help unravel some more of that complexity.

But a good starting point for greater understanding, is trying to throw off the prejudices of the past. For a long time 'the brain is like a computer' analogy was so overworked, that many researchers in the area didn't want to hear that idea mentioned again.

Yet it was a clue, and in the end most researchers headed off to see what kind of *software* solution could be found to understanding the brain.

They began looking *outwards*, exploring software-based algorithms that learn to 'think' (artificial intelligence); or mimicking the brain's structure in software (development of neurological networks, now rebadged as 'deep learning'), which are both effectively kinds of software-based research.

The idea of also looking inwards, and seeing DNA coding as software, (and how that changes), is another starting point for new areas of research on the brain. It is an approach that starts with DNA code, starts studying epigenetic changes as something like 'switches' in code, and sees how it all fits together & evolves.

It is an idea that was always *hidden in plain sight*, based on the idea that the brain is not a computer; it is software, biological software. This new world of *biological coding* is only just beginning, compared to the world of digital coding. Can we learn to code DNA to a level of sophistication needed to achieve meaningful results in making sense of the brain? Not even close, but it is another entry point to greater understanding.So, while we just don't really know how the brain works yet, we have some good ideas. Software that runs the brain is *not* the kind of software that exists anywhere else right now. It's a wonder of nature. It is the most complex piece of software we know of in the universe. And it is built from DNA code, and evolutionary impacts on that core software.

In summary, the idea that the *'brain is biological software'* is not just a change in definitions, it is a change in how we all see the brain. Its mysteries can be now seen as a *software riddle* to be solved, rather than an impossible equation of unknown dimensions.

So, the brain's *complexity* is not vanishing. The more about it we know, the less we seem to understand, and the more we are in awe of its power and sophistication. We are still grappling to find a framework that can explain it.

What is *vanishing* is the idea that you can understand the brain without seeing it as software.

Once we accept that the brain is a kind of *biological software,* the road to fully understanding it will become clearer. It's software of a sophistication well beyond us now. But maybe not forever, as the brain's software mysteries start to unravel with more research into the future.

5. Rethinking Biological Profiles

So the human brain remains largely a mystery. But we need to try to understand some key kinds of behaviour in all living things, before we can really understand *life* as we know it now – and what it could be in the future.

Intelligence, Sentience & Consciousness are central to the behaviour of all living organisms. But the meanings of these 3 fundamental attributes are often left undefined, are vaguely defined, or are assumed. Below is an effort to rethink what they mean.

It will open a window on how they could work in future non-biological life forms.

Introducing Intelligence

'Intelligence' is very hard to define. Despite it being a topic of high interest, there is still no consensus about how to define it or accurately measure it. Is it a general ability, or a set of skills and aptitudes, like seeing and problem solving, that takes logic and reasoning?

It took a while for people to start thinking about intelligence. Aristotle, the well-known Greek philosopher from around 2,300 years ago, discussed 'reason' – an early reference to something close to 'intelligence' – as our ability to control our (human) emotions. Then in the 19th century Darwin concluded that more 'evolved' species had more mental powers. It was really in the 20th century that research on intelligence picked up speed.

The word 'Intelligence' is of course defined in several dictionaries. The Cambridge dictionary defines 'intelligence' as the ability to learn, understand, and make judgements or have opinions that are based on reason.[106] The Collins dictionary defines 'intelligence' as the ability to think, reason and understand instead of doing things automatically or by instinct.[107]

These are both very human-centric definitions – humans have 'opinions'and ability to 'understand'. It's a starting point.

There are more flexible definitions going around - like Yuste & Levin's definition describing 'intelligence' as *'a purposeful response to available information'*[108]. It means that any creature that reacts to their environment or situation or a problem - in a purposeful way - has a form of intelligence – including single cells or even viruses. That is a minimalist definition, but at the same time, it does *not* underestimate the intelligence of many living things.

For instance, many creatures can solve problems, and make decisions that help them; and that is the essence of intelligence that leads to survival and then reproduction. Whether it is the crab stuck in an evaporating pool that somehow climbs out, and moves to a bigger pool with more protection from the sun; or a bird that seeks shelter in a storm. Decisions need to be made, and problem solving needs to be done, by a wide range of creatures. Just how the decision is taken based on certain inputs, is interesting, and we will return to that later.

This flexible kind of understanding of 'intelligence' starts to bring back a lot of behaviour of many living things into the definition of 'intelligent' behaviour.

It's taken a long time to escape the shackles of human-centric views of the intelligence of other creatures.

For instance, *octopuses* have long been known to exhibit a lot of playful and creative behaviour; and perform badly in some scientific tests because they don't seem interested in them, or even try to 'get around' them. They were the creatures that led to some researchers into animal intelligence like Peter Godfrey Smith to say that 'there is no single scale on which intelligence can be sensibly measured'.[109]

To Peter Godfrey Smith, the Octopus is curious, flexible, adventurous and opportunistic[110]. They find pairs of half coconut shells (cut and discarded by humans) and carry them around as portable shelters, or reassemble into a sphere with itself inside, seemingly just experimenting, having what we would call 'fun'. It's a kind of intelligence in the past not understood.

Experiments in the 1950's to 'train' Octopuses to pull a lever for a reward, found that they would not always do it. Once they broke the lever they had to pull, and another time pulled a light into the tank that was part of the experiment, and often directed jets of water out of the tank, in the direction of a particular experimenter. The behaviour interfered with the conduct of the 'experiments', naturally, and left the researchers thinking their experiment had failed in some way – the variables could not be explained.[111]

Louis Lefebvre from McGill University invented the first scale of intelligence for birds more than 20 years ago, based on birds' behaviour in the wild.[112] He looked at how inventive the species was in the natural environment, how they make and use new things, and how they find creative solutions to problems.

These things are indicators of behavioral flexibility, which author Jennifer Ackerman sees are a fairly reliable measure of intelligence. The ability to do something new, and change behaviour.[113] It's easy to see the echoes of some Octopus behaviour here just mentioned above, sometimes wrongly concluded as a sign of them being bad test subjects or lacking intelligence.

Lefebvre found around 2,000 reports in journals over the past 75 years describing innovative behaviour in birds of many species. Like the gull regularly hitching a ride on a garbage truck to an organic composting facility west of the San Francisco Bay.[114] Anecdotal, but happening all the time, a bird discovering a better way to find the next meal. Examples of birds solving problems using cognitive skills not instinct or conditioning, are fascinating, especially since a typical bird brain is the size of a walnut. Part of the explanation of this behaviour is that the neurons in bird brains are smaller and more numerous and densely packed than those in mammals.[115]

Researchers like Sue Healey of the University of St Andrews are now investigating what birds can remember in the field, adding new layers of knowledge about birds' behaviour & intelligence – like the fact that Blue Tits build different nest structures to deal with different temperature.[116] Or the hummingbirds who remember almost everything about the flowers they visit, or the Vasa parrots and Goffin's cockatoos that make and use their own tools.[117]

Birds can make tools, use language-like communications, and reasoning[118], that are all signs of intelligence, similar abilities to humans, yet for a long time, to attribute these things to birds as not seen as correct, as humans were the only ones with these skills.

Which brings us to the Monarch butterfly. A female Monarch butterfly born in Boston will migrate to Mexico, in one of the greatest migrations of the natural world. Biologist Douglas Blackiston, the entomologist Adriana Briscoe, and some other colleagues tested monarchs, with fascinating results.[119] They trained to butterflies to change colors to find a sugar reward. Rewards were associated with yellow and blue and red. Once they realized they would get a reward if they changed colors, that's what most did. Blackiston felt that the prime characteristic of the monarch butterflies is intelligence – their brains allow them to learn.[120]

As the millions of Monarch butterflies fly south from the USA to their favorite mountains in Mexico, they form a cloud of insects that seem 'like a veritable river of flying colors, glittering in the sunshine'.[121]

But with the changes to the places they travel over, they do need to be adaptable, so if their travel plans were 'hard-wired' in their brains, and they couldn't adapt their plans, they might all die out in some disaster. An example of adaptation is how they get across the Gulf of Mexico. There are so many ships in the Gulf now, that many Monarchs use a whole new route – hopping from ship to ship out to sea. Using nice resting places on the journey.[122]

So reminiscent of the gull regularly hitching a ride on a garbage truck to an organic composting facility west of the San Francisco Bay.

In addition to birds, many insects display an amazing range of intelligence. There are more than 200 million insects for every human being living on planet earth[123], according to Ann Sverdrup-Thygeson, in her book 'Extraordinary Insects'.

Before humans existed, insects were involved in agriculture, (termites grow fungus for food) and animal husbandry (white ants keep aphids as dairy cattle), as well as hunting (caddishfly larvae catching other creatures in net-like webs). They are simply the most successful class of animal on the planet – by number.[124]

So are insects intelligent? Again it depends on the definition, and Ann Sverdrup-Thygeson's clear answer is 'Yes', if you define intelligence as the ability to acquire and analyze information, as well as learn and make judgements. That's quite a high bar. She acknowledges that intelligence varies between insects; but finds the most intelligent insects look for food in lots of different places, and also form close bonds with one another. So social insects like ants do well on this criteria – they divide up jobs, share experiences, and communicate in an advanced way. And they can train other ants[125]– an interesting one, since the ability to teach was often seen as special to humans.

More and more aspects of insect intelligence are being discovered all the time, building up a complex web of behaviour that is not inherited from birth. Karl van Frisch won the Nobel prize in 1973 for discovering that bees can see colors; and can tell each other where to find food by performing a sophisticated dance[126].

There is a *language of dance*, in which the honeybee returning to the hive tells other bees exactly where the pollen was to be found. The bee wags her body from side to side, then does a half circle to the left, then another straight waggle, then circles to the right, then repeated. From that other bees know where to find the food. If the food is in the direction of the sun, the dancer will waggle straight up the honeycomb. If it is 45° to the right of the sun's direction, the waggle will go 45° up to the right, and so on. How long the waggle goes for tells other bees the distance from the hive.[127]

That's not all instinct, it is skillful – and intelligent, and features spatial memory (memory of locations), and clear communication with other bees. It may have had an evolutionary basis in the need to know where their 'home' (hive) was in relation to food they were seeking. [128]Insightful author Lars Chittka says that many common activities like 'wax comb' building, are rarely pre-programmed, and need to be partially learned, and involve high degrees of flexibility and even planning skills. [129]As she says; instinct interfaces with memory and cognition.

So there is lot of evidence of bees' intelligence, and growing awareness of this as research continues. But author Lars also thinks bees can be defined as being *conscious*; able to feel pain and show signs of other emotions (like the excitement of swarming); and seem to have the ability to 'know what they know'.

But she also acknowledges that while there has been no formal proof of consciousness of any animal, but there is a growing recognition of it being possible as more research piles up. The human-centric approach is starting to drop away. Lars proposed that bees' consciousness would be a simple form, and different to that of humans.

This is a great point, that consciousness can be different in each species, but also very real. It led Lars to say that *comparative consciousness* is an idea and budding discipline just beginning. [130]This author is running into the need to rethink consciousness and how it relates to different species. We will return to this idea later.

The world of fish is another massive area of evidence. For instance injured tiny zebrafish are known to switch from a natural environment to a tank suffused with analgesia to relieve pain, [131]suggesting some consciousness - as well as intelligence.

Finally, there is a growing pile of evidence about the behaviour of farm animals, revealing intelligence and sentience across a wide range of farm animals. Who would have thought that chickens are capable of complex communications, or that cows grieve when their calves are taken away from them?[132] It's true.

Animal intelligence – whether in birds, insects, fish, octopuses butterfly, or farm animals; is not something fully researched or understood; but great researchers all around the world are continuing the process of finding out more. And hopefully bringing less and less 'human-centric' thinking to the topics of research.

So over time, in scientific studies and general society; *'intelligence'* has emerged as something we see as important, even if we can't fully define or seriously measure it. *'Intelligence'* unfortunately still remains a 'place' for ranking humans against humans, and humans against other living things.

IQ & EQ

Two more terms need to be reviewed – IQ (intelligent quotient) and EQ (emotional quotient).

IQ means 'intelligent quotient'. It is best described as a measure of academic abilities. It is usually shown as a standard score, that tells you where sit in your peer group, out of an average of 100 points.

It is one of today's modern mix-ups to see 'intelligence' as the same as 'IQ' (intelligence quotient). IQ was designed as a way of measuring some aspects of intelligence for humans, so it is a much narrower term than 'intelligence' in general.

The term IQ was originally conceived by Alfred Binet in 1903 as an 'intelligence test', it is now in widespread use in schools, for jobs, and many other applications.

Edwin Boring picked up these ideas and developed them for use with *intelligence* testing of soldiers during World War I in the period 1914-1918, and then in subsequent work. He later became a Professor of Psychology at Harvard University.

He was concerned about the scientific objectivity of this, and in later life, when asked about his definition of intelligence, said "Intelligence is what the tests test".[133]

Widespread use of IQ tests by large corporations ended up missing some talented people, and still does. Just because *intelligence* is not necessarily measured in your IQ result.

For instance, Microsoft used to ask applicants for a job at the company a whole suite of questions to test their intelligence, including the classic 'Why are manhole covers round'?[134] If you didn't know the answer then you might not get a job at Microsoft. Really? By the way, the answer was that manhole covers are round so that the cover can't fall into the manhole, even if you turn them, whereas if you turn a rectangular cover on its side it can fall in[135].

However, it seems like the relationship between IQ and intelligence could be more complicated that it seems. IQ tests are tracking a certain type of intelligence, a mix of math and facts and other general knowledge. It's a narrow view of intelligence. So it could be possible to be a human with low intelligence to get a high IQ test score.

Also, how does the relationship between intelligence and IQ work the higher the IQ measurement is? It looks like it works only up to a point. Someone with an IQ below 70 is seen as mentally disabled, 100 is average, and just above that will get you into college or university. But once IQ gets to 120 or more, some research indicates that it does not seem to translate into any measurable practical advantage. [136]Still most people would rather have an IQ of 120 or more, than 100, or less.

Today IQ (intelligence quotient) and IQ Tests are part of modern life, but if we take Edwin Boring's concerns about objectivity on board, (as noted above), then the main question we can ask about IQ Tests; is 'Do they measure anything else apart from how well you do in IQ Tests?' The answer is 'No, probably not'.

Later, when we look at artificial intelligence (AI), it can do very well in IQ tests, and famously wins quiz shows and passes law exams – especially when it is trained to know what to expect. But is the AI tool intelligent? We will look at that later, but the initial answer at this stage is 'No'. After all, if the AI is simply regurgitating facts that it has been trained to learn, it does not tell us much about the intelligence of the AI, but it does tell us how well trained it is.

EQ (emotional quotient) is another measure of intelligence. Often called Emotional Intelligence, EQ is defined by the Cambridge Dictionary as 'the ability to understand the way people feel and react and to use this skill to make good judgements and to avoid or solve problems.'[137] Emotional Intelligence can also defined more simply as "the ability to understand, use, and manage our feelings".[138]

The other point of interest with EQ, is the question of whether you need to have emotions to have a high EQ. We found that you don't need much intelligence to have a high IQ, as tests don't necessarily tell you much. But as far as EQ is concerned, does the same logic apply?

The answer is 'Yes', you can be a person with few emotions, who can simulate empathy, concern and the whole range of human emotions. This will apply to *sociopaths*, who lack empathy but can pretend they have it through lying and manipulation. But is also true of computer software *simulating* emotional responses based on coding of facial expressions. The software is showing a high EQ, but almost certainly does not have any emotions in its own right. This kind of intelligence is reviewed later in more detail.

IQ and EQ are not opposing competencies but separate yet related ones. We use both in many day-to-day interactions. They are part of our understanding of 'intelligence'. High IQ can be accompanied by low EQ, but this is more like a stereotype than a common occurrence[139]. For example, the technology 'geek' who can't *'do'* conversation with people.

So IQ can most likely predict how you will go academically, but there are other types of intelligence like EQ, that are just as important. In addition, there are other types of intelligence like creativity, insight and 'common sense' that are very hard to *pigeon-hole* or define easily.

And there's more types of intelligence that those mentioned above. Russell Kenwick in his book on *'Understanding Your Emotional Intelligence'* noted that that spatial intelligence is the ability to think in 3D, musical intelligence is the ability to perceive mood, rhythm and tone, while there is also athletic, artistic and mechanical abilities that are different types of intelligence[140]. So there is more than IQ or EQ to intelligence.

The wider the definition of intelligence is set, the more types of intelligence are found. And the more creatures can be defined as having intelligence of some kind. We don't need to hang onto *'human-centric'* views of intelligence. That old view of intelligence is *vanishing*.

Introducing Sentience

An important related term to 'intelligence' is *'sentience'*. The Cambridge dictionary defines this as meaning the quality of being able to experience feelings.[141] The Collins dictionary gives sentience the meaning of 'being sentient', with 'sentient' meaning a sentient being is capable of experiencing things through its senses.[142] Sentience is an important link the chain of ideas here.

The words 'sentient' and 'sentience', as Nicholas Humphry notes[143], came from the 17th century to describe any creature – human or otherwise – that responds to sensory stimuli.

But over time (until recently) 'sentience' narrowed down in meaning to emphasize what the sensations '*feel like*' to the subject, and also ask whether animals have experiences that feel to them, the way ours do to us.[144] It became about the quality of the sensations or 'qualia'. Humans can feel the pain of a bee sting or the squinting of eyes in the bright sun, or the feeling of sadness on the death of a partner.

So do all other creatures feel things the way humans do? Unlikely given the differences between many creatures. But does that make those feelings any less meaningful to other creatures? For example, many creatures are well known for showing grieving-style behaviour at the loss of a partner - elephants mourn their dead,[145] or the loss of a friend.[146] Probably more than some humans view the death of a relative. So what scale are emotions on?

Feeling pain or pleasure, joy or disappointment, cold or warmth, fear or excitement, thirst and hunger; are a big part of the way many living things respond to the environment and other living creatures.

This idea that if a sensation doesn't have much depth or quality to it, then it somehow matters less, is a truly strange idea. Somehow there is a 'hidden' scale that says a primitive organism will have less awareness and less depth of experience of, say, pain, and so it matters less. This imaginary scale might be fine at the extreme ends, but when you come to a wide range of living creatures, from fish, to reptiles, mammals, and try to separate out their ability to experience pain or happiness or hunger, or the quality of that experience, it starts to become absurd.

A lot of scientific experiments (especially in the 20th century) on creatures to assess their response to rewards and punishments, uncovered more than they realized at the time. Often they were seeking to test 'intelligence', using another imaginary scale.

But in searching for *intelligence*, they sometimes forgot about sentience. So we hear tales of some octopus experiments from 1950's in the Naples Zoological Station, that dished out electric shocks (negative reinforcement), and removed parts of the octopus's brain, or cutting nerves, to see what would happen when the octopus woke up – usually after an operation without anesthetics.[147]

Reading of this behaviour of some scientists from the 1950's would lead many to wonder which of the 2 creatures – the octopus or humans - were more intelligent? It is very hard to answer and shines a light on how we can really define 'intelligence'. Where does that *entitled* human behaviour fit in? Where does that lack of awareness of other creatures' pain and suffering come from?

It's mainly derived from the idea that *only or mainly* humans can fully experience these feelings to their full extent/quality, and everything else experienced by other creatures is somehow less. This is the idea that we are actually projecting 'human' emotions onto other creatures.

It is called *anthropomorphism,* attributing 'human' characteristics to an animal. This view is that when we, as Homo Sapiens, talk about dogs in pain or starving zebras or injured bees; we are really just *projecting* 'human emotions' onto other creatures *(anthropomorphism)*, when actually these other creatures are not really experiencing those actual emotions – as we know them.

This concept has been the excuse & justification for much animal suffering, and ill-treatment. Now is the time to start rethinking that whole idea.

This kind of argument – that some creatures are not really 'sentient' as we are just projecting human emotions onto their actions - still carries great weight in a world where Homo Sapiens regularly destroy other environments and routinely kill other creatures as part of our own 'human' plans.

However, many more modern scientists are aware that things like pain are very subjective, and it is impossible to measure suffering objectively, or even assess it in anyone other than yourself.[148] The fact that a dog with an injured leg can howl, and display suffering, carries more weight to some, compared to how insects manage pain. Added to that, is the fact that some creatures are able to modulate their pain response, further masking the level of pain.

So we need to rethink *'sentience'*. We still have a lot to learn about interpreting the actions and behaviour of other living things, including the evolution of social behaviour in relation to all creatures.[149] It's OK to acknowledge sentience in other creatures, and that does not diminish us, it lifts us up, to start seeing our human actions for what they are - in some cases affecting or hurting other living things.

So-called primitive creatures like jellyfish receive stimulus and then respond. It's not a matter of being trained to respond like Pavlov's dogs, the adaptive responses are part of the cellular basis of behaviour. For the learnings to be long-lived, a nervous system is needed.[150] Likewise insects like ants can learn, remember, think and communicate.[151] It is more than reflexes.

Taking a tougher line, Jonathan Birch, an associate professor of philosophy at London School of Economics, argues that reflexes are not enough to establish sentience, as an animal has to be shown to have a central system that values and disvalues stimuli.[152] Though that in itself may be a very "human" view of things.

The main supposed peril of *anthropomorphism* is the tendency to wrongly assign human emotions or mental abilities to other animals. But it often sounds like a way to 'define away' similar experiences (like pain) in many animals – after all, the argument goes, it is not really (human) pain. It is a very shallow approach to understanding all creatures. There is something wrong with a human-centric approach to these questions.

And author Ed Yong argued that there is a tendency to frame animals' lives in terms of *our* senses rather *than* theirs.[153] The classic example of this was in Thomas Nagel's famous 1974 essay 'What Is It Like to Be a Bat?'[154] To do that, you have to imagine yourself with webbing on your arms, eating insects and flying via sonar. Can any human truly know that?

In relation to human behaviour towards animals though, of the 3 concepts (out of 'intelligence', 'consciousness' and 'sentience'); *'sentience'* is the idea on the front line of human treatment of other creatures.

Human behaviour is starting to be a trigger point for new legislation[155] to help protect other living creatures from many forms of human behaviour, especially behaviour that inflicts pain on living things who do in fact 'feel' it.

Cruelty to many animals has gone on for way too long – whether it happens in training animals for the circus, conducting cruel scientific experiments, animal farming of many kinds, animal-based sports of many kinds, down to horrific & cruel bearbaiting or dog fights still going on today in some backward places. Cruelty to other living things also extends to constant assaults on the natural environment to extend cropping by mowing down forests, affecting both wildlife and plants.

It is a gigantic area of human behaviour that has escaped systematic treatment by the law in many places. Mainly because many humans do not see anything 'wrong' with most of the behaviour described above.

Eventually we may start to see the unnecessarily cruel parts of this behaviour as equivalent to doing this kind of behaviour to other humans. But for now, efforts of researchers like Jonathan Birch in the UK, (and his 'Foundations of Animal Sentience' project), will hopefully start to clarify sentience in many living creatures including insects; and new animal welfare laws will start to change the way we think and the way we treat animals.

Introducing Consciousness

The other big concept that needs exploring is *'consciousness'*. What exactly it is, and how it arose and works, has eluded scholars for generations, yet the word has a widely understood meaning in popular culture.

The Cambridge Dictionary defines it as the state of being awake, thinking, and knowing what is happening around you.[156] While the Collins Dictionary defines it as your mind and your thoughts.[157] The definitions of 'consciousness' here are also very vague and general, but are a starting point.

Nicholas Humphries' great definition says *consciousness* means having knowledge of what's in your mind - your conscious mental states. They could include memories, emotions, thoughts, wishes etc.[158]

This idea of consciousness of a 'sense of self' is clear, in that is all about the fact that you know what's in your mind – now, in the past, and with ideas about the future.

Another word is worth mentioning here – *awareness.* It is defined as "knowledge that something exists, or understanding of a situation or subject at the present time based on information or experience".[159] *Awareness* is usually seen as being synonymous with consciousness. But it is a lighter and more general term.

As far as *consciousness* is concerned though, virtually all researchers wonder where it comes from. What physical structures in the brain create it? If the mind and consciousness are actually forms of *information processing* implemented by the brain, then maybe biological brains are not needed, and consciousness can be created on a computer? Many authors, including Anil Seth, doubt if something like consciousness can be simulated on a computer.[160] But it is also an area of great potential.

It is great question to ask, to consider whether consciousness can be created on a computer. Amongst other things, it does depend on the type of software and type of computer. Nothing we have now can achieve consciousness.

There is enormous research going on in this area, as there is so much we don't understand. For instance, many living creatures were previously thought NOT to have consciousness, are being studied more closely.

So our understanding of consciousness is just starting to come under challenge, as now it is really only a human-centric understanding. The definition needs to widen to make sense. It's not purely the preserve of Homo sapiens.

To get a *helicopter view* of consciousness we are going to compare it to sentience and intelligence in the sections below.

Sentience & Consciousness Compared

Some authors like Nicholas Humphrey define 'sentience' as a form of consciousness – *phenomenal* consciousness[161], as opposed to cognitive consciousness. It basically argues that 'sensations', (touch, vision, sounds etc.) and a creature's response to these are *phenomenal* consciousness, and if a creature does not 'experience' or 'feel' these things, then they are *not* sentient. Whereas cognitive consciousness is a higher level of consciousness involving awareness of being - you know who or what you are and have a past, present & future as 'I'.

This idea of drawing an artificial line between creatures – a distinction between insentient and sentient animals – based on the idea that consciousness is either 'on or off', does raise obvious questions. Humphreys ends up arguing that phenomenal consciousness is restricted to warm-blooded animals – all mammals and birds.[162] So insects, fish, and all reptiles get left out – including snakes, turtles, alligators, lizards etc. And most controversially, even octopuses get left out.

Other authors like Peter Godfrey-Smith, who studied octopuses for years noted in his book *Metazoa*, that after 10 years of studying octopuses, that their behaviour left him in no doubt that octopuses experience their lives and are conscious in the broad sense of that term.[163]

They show complex behaviour, like solving puzzles, and being playful, and exhibit what some would call lateral thinking, in disrupting experiments on them. To put them outside an imaginary line called (lacking) phenomenal consciousness, yet at the same time acknowledge that they may be cognitively consciousness,[164] is not really clarifying anything.

An important milestone happened on 19th April 2024 when around 40 scientists signed The New York Declaration on Animal Consciousness at a conference in NY. The declaration called for formal recognition that there is strong scientific support for attributions of conscious experience in mammals, birds, possibly fish, amphibians, reptiles, and many invertebrates like crustaceans and insects.[165]

The scientific world is starting to bring to the attention of the human world in general, that consciousness is not unique to Homo sapiens, no matter how long some have tried to deny it. Many creatures recognize themselves in the mirror, (an indicator of consciousness), have social structures, use some language, and use tools. They can 'feel' pain or fear, and experience tastes and smells. All to differing extents, and in different styles.

So there are obvious problems in drawing an artificial line between creatures, and placing them on each side of the *line* based on some assessment of sentience and/or consciousness. It is just too arbitrary.

It could be true that having the concept of 'sentience' as being part of 'consciousness' makes some sense. Many creatures 'feel' things that we would call pain, and it is the natural & efficient working of their sensory systems. And it implies some sense of *self*, what Humphries called 'cognitive consciousness'.

But it could equally be true that 'sentience' should be seen as separate from 'consciousness', as it tends to be, in day-to-day usage of the words. After all, it is hard to know & define exactly what level of *sense of self* that each creature has. In the end, where a creature responds to stimulus, and it is not a simple reflex, you have sentience. It really is hard to dream up all sorts of rules to define how a creature has some 'sense of self'(consciousness).

Why should we not consider the possibility of sentience without consciousness? To some extent, defining consciousness generally as 'experience' that includes reactions and behaviour (eg. reaction to pain) we would typically call *sentience*, does devalue the meaning of consciousness. That approach may be as bad as artificially splitting creatures into sentient and non-sentient.

For instance the spider who has lost its leg in a fall, or the ant half squashed by a human foot and barely alive, or the butterfly captured by a child half-way through its only day of life; would be aware (to their own) extent of their predicament. Is there fear? Or pain?

All we know for sure is that at some level they know there is a problem, and that is *sentience*. Is their version of pain or fear different from ours? Yes, in quality terms you could argue that ours is a deeper experience based on more knowledge and pain receptors. But can we really *define away or discount* the quality of other creatures' reactions and responses as 'inferior' to ours?

With millions of different, complex creatures in the world, is the discussion of sentience really so 'black and white'? Nobody is fully sure of what *sentience* is, and where it starts and stops.

For instance, in the movie Mission Impossible 7, that came out in 2023, a software 'entity' became *'sentient'*. By sentient they probably meant not only *conscious*, and aware of its own identity, and able to think for itself. But we can probably deduce from its behaviour, in wanting to take over the world, that it was also sentient, in that it was able to feel emotions.

In Nicholas Humphrey's terms, they meant the software had developed *cognitive* consciousness, where it knew what it was, and had a past, present and future as 'I'. As well as what he called phenomenal consciousness – like touch, vision and sounds and emotions.

So the term *sentience* has drifted and expanded (in some parts of popular culture) to mean conscious. Language is a changing thing, and this is no problem. But it is handy to remember that sentience has a narrower meaning that can apply to all creatures, referring to ability to feel emotions.

This is especially important in understanding creatures who can feel heat or cold or pain, and may lack the higher level of consciousness, like the extremophiles or viruses mentioned in Chapter 1.

There is no need to draw strict artificial lines around these terms like sentience and consciousness, no need to see only 'black and white'. Otherwise, some creatures get defined as not being sentient, or not being conscious, for no good reason.

There is a need to see past, and cut through, the current rubbery definitions of intelligence and sentience, to create a more rounded view of each creature's standing. A solution to this problem is offered later in this Chapter.

Intelligence & Consciousness Compared

A great work in the area of understanding 'intelligence' and consciousness' came in 2021 from Anil Seth, one of the world's leading consciousness researchers, with his work called 'Being You',[166] that actively and thoughtfully considered and compared intelligence and consciousness

In many ways, by comparing them, he shed more light on them, and broke down the meaning to something that we could get hold of and place into a logical framework.

He saw consciousness as more to do with being *alive* than with being *intelligent*, in that 'selfhood' is a deeply embodied biological process that underpins the simple feeling of being alive; and is the basis for all our experiences of 'self'.[167] Intelligence is another kind of process not necessarily related.

He also saw conscious *experiences* of the world and self are 'forms of brain-based prediction,'[168] as the brain constructs the world around us and our place in it based on built up memories and sets of knowledge. But just how we get these *experiences*, (eg. feel an emotion or entertain a mental image),[169] derived from our physical systems, is the hard question.

This issue was described as the 'hard problem of consciousness', a term proposed by Professor (of Philosophy and Neural Science) David Chambers from New York University, in the early 1990's.[170] It is the problem of how and why physical processes give rise to consciousness.[171]

David Chambers did not see any solutions to explaining consciousness at the moment that have any consensus, but there are a lot of theories, from denying consciousness, to arguing it is epiphenomenal, (an incidental effect in the brain of electrical or chemical activity) or illusionism (an introspective illusion).

Introducing Profiles

The other topic that interests researchers in the study of the brain is how to 'measure' consciousness. The debate was long dominated by the approach of building and defining 'levels' or a 'scale' of consciousness, with the top level being the human species. That idea of humans on top of the ladder was really based on historical views of the importance of Homo sapiens, which can now be seen as open to challenge.

An alternative idea emerged in 2017, from the work of Jonathan Birch, Alexandra Schnell & Nicola Clayton; the view that we could see the dimensions of consciousness as *'profiles'*.[172] They came up with the idea that each species could have its own distinctive consciousness profile.[173]

Dimensions of consciousness include perceptual richness, evaluative richness, integration at a time, integration across time, and self-consciousness.

These dimensions determine the extent to which an entity can perceive, evaluate things at the time, and over a time period, and the extent to which they recognise themselves as a separate entity.

Other leading thinkers like Tim Bayne, Jakob Hohwy & Adrian Owen also recognised that the idea of 'levels of consciousness' had become untenable, and a more 'multidimensional account of global states' was needed.[174]

The beauty of this idea of 'consciousness profiles' is that each species can have its own profile, and there is no single scale (or levels) along which species can be ranked as more of less conscious. Instead of that, each species has its own distinctive consciousness profile.[175]

We can start to see the beginning of an emerging science of animal - and plant – consciousness; that allows for all living things to be studied and understood in a broad framework, that is not based on what we understand as 'human' consciousness.

It immediately lets us start thinking of many other creatures that we have known for some time to have complex brains & behaviors - like octopuses for instance, and gives us the opportunity to place them in a much better and more sensible framework than in the past.

Extending the idea of Profiles

But beyond the immediate application of the idea of 'profiles' of consciousness; it is clear to this author that this 'profile' concept should be brought into the spotlight, and can also be used to understand, redefine and rethink both *intelligence* and *sentience*.

'Intelligence profiles' are a real way to completely rethink what we mean by 'intelligence'. And 'sentience profiles' are a real way to completely rethink 'sentience'. Making this approach the *key* to understanding these ideas of consciousness, intelligence and sentience is a new way of imagining and analyzing this whole area. *Profiles* sit on the edge of the discussion now. They should be much more than this.

So every living thing in reality actually has an *intelligence* profile, a *sentience* profile, and a *consciousness* profile. These things differ for all creatures. It's an idea that is self-evident, yet somehow the significance of the idea has never been fully explored.

The idea of a *single* 'ladder' of sentience, or intelligence, and of consciousness, can be returned to the dustbin of history as a simple idea that reached its used-by date.

Suddenly the *single scale* has gone, and each creature can sit in a framework of profiles that allow us to see it as it really is – special.

We can still compare species in some meaningful ways – say by outcomes of actions. It's not saying that one consciousness profile, for example, is higher or better than another consciousness profile. But outcomes can be used to consider if the consciousness profile is affected by other factors in the sentience profile, like aggression, for instance.

We are saying that consciousness consists of (in Nicholas Humphrey's terms) cognitive consciousness, (where a creature knows what it is, and had a past, present and future as 'I'), as well as what he called phenomenal consciousness – like touch, vision and sounds and emotions – essentially what we can sentience.

So the consciousness profile could reflect the intelligence of a creature, as well as its emotional nature, plus some other 'X' factors like knowledge, experience, common sense and wisdom, that is really going to change the helicopter view of that creature. It is a more *rounded view* of consciousness. Similarly using profiling will give great rounded views of intelligence and sentience when they are considered separately.

Losing that idea of an implied single scale of *consciousness* or *intelligence* or *sentience* does not invalidate a large body of scientific work done over the past few centuries. But it makes us want to look for any assumptions in that work.

For instance, the original work that *classifies living things* as begun by Carl Linnaeus, whose 1735 work *Systema Naturae* (1735), and more recent updated versions, created a system of 'taxonomic' classification, a modern system of naming organisms. The original work was only 12 pages long. But by the 10th Edition in 1758, it had been expanded to include 4,400 animals and 7,700 plants.

Successive researchers have taken the study and definition of all living creatures much further. But we don't need to rank them into one scale, or even see them like that, in some kind of crazy ladder that has humans on top of that ladder.

In comparing consciousness, intelligence and sentience, a few observations can be made. *Sentience* is the least understood, and the least respected, yet in many ways the most important. If a creature is capable of experiencing things through its senses – whether pain, fear or excitement, does it have any less rights than Homo sapiens?

Which means the octopus can be seen for what it is. A unique and special creature, with blue blood, three hearts, and 8 arms that each contain a 'mini-brain', able to act independently, and taste, touch and move without direction. Having 9 brains gives them about 500 million neurons, as many as a dog, allowing a lot of problem solving and sneaky fun behaviour too.[176]

It is intelligent, sentient and conscious in the broad sense of the terms. It feels things and is aware of its own existence. And its sentience, intelligence and consciousness profile is one of the most amazing ones you could ever find. We don't need to leave it on the outside of any definitions anymore, it is what it is. Which is a fascinating creature.

Can't See the Forest for the Trees

Beyond freeing up definitions and understanding of living creatures, profiles start to shine a light on our own (human) behaviour.

There's a whole emerging area of study reflecting on human species domination of other species, including humans transforming the environment, which has changed, and continues to change, the planet forever – with machines as our tools of destruction and creation.

It is sometimes said that humans *'can't see the forest for the trees'.* We see many separate complex issues but cannot grasp the overall problem. We sometimes follow the detail but miss the 'big picture'.

Climate change used to be like that, but more and more scientists and people on the street are starting to see it as their problem too.

Human induced climate change and global warming are just the latest in a long line of human interventions in planet life without consideration of the *sentience* and rights of other living creatures that we have shared the world with for a long time.

If sentience is under-respected, then *intelligence*, on the other hand, is over-respected, misunderstood, hard to measure, and in some ways less important than sentience & consciousness.

'Intelligence' did become an *anthropocentric* (human-centric) idea, in that other species are often seen as having 'less' intelligence...even if we are not sure what 'intelligence' is.

Merlin Sheldrake in his book *'Entangled Life – How Fungi Makes our Worlds, Change Our Minds, and Shape Our Futures'*[177] is one of the breakout thinkers in this area. He notes that classical science defines 'intelligence' by using humans as the yardstick by which other species are seen.

In the past, humans were put at the top of the rankings, then you move *'downwards'* in a league table, to 'less intelligent' species.

Yet when seen through the prism of profiles, it is legitimate to ask some different & hard questions about the intelligence, sentience & *consciousness profile of species* – whether humans in human society or trees in the forest.

But by adding an intelligence *profile* of humans, for instance, onto other aspects of human behaviour like our emotional profile; intelligence can become part of a wider consciousness profile, which reveals that human intelligence in reality is far lower than many think, due to the power of our emotions, as evidenced by the destructive outcomes in successive conflicts and wars.

But if that is the story for *humans*, what of *forests*; what is their *consciousness, sentience and intelligence profile*? It's not thinking about a single scale, but about looking at behaviour and outcomes from different angles.

To understand this more, we need to go no further than the work of Suzanne Simard, Professor of Forest Ecology in the University of British Columbia.

Her research on tree connectivity and communication, and its effect on forest health and biodiversity, has given her global recognition for her pioneering work. Her work from 1997 today, includes the amazing work, 'Finding the Mother Tree – Uncovering the Wisdom and Intelligence of the Forest' is a fascinating insight into the 'language' of trees and 'intelligence' of the forest.[178]

Her work was first published in 1997 by Nature and called the 'wood wide web', detailing below-the-ground communication between trees via a community of mycorrhizal fungal species that shocked and surprised many, and still has not been absorbed by general society. Trees were communicating via a web.[179]

She discovered the trees in a forest have a web of interdependence, linked to a system of underground channels, often featuring fungi and roots, where they 'perceive and connect and relate with an ancient intricacy and wisdom'.[180] In short, trees and plants can be part of an interconnected ecosystem.

Via hundreds of experiments, she uncovered tree-to-tree communication, and relationships that were initially highly controversial, but now peer reviewed and widely published.

And interestingly, for organisms that are still not seen as having a 'brain', they act in a way to support seedlings, and help struggling trees. In the forest a fungal network can pervade an entire forest, much like synapse and nodes in the human brain. Trees can respond to each other by emitting chemical signals too. Again, much like our neurotransmitters. Suzanne Simard describes it as signals created by ions cascading across fungal membranes.[181]

So signals can be warnings, recognition or signals of distress where help is needed. In her view, the forest is wired for wisdom, sentience and healing'.[182]

Other authors like Peter Wohlleben in 'The Hidden Life of Trees'[183] have backed up Suzanne Simard's work and explored it further.

Peter Wohlleben found that trees can share food, even nourishing the stump of a felled tree for centuries after it was cut down, by feeding it sugar & other nutrients, so keeping it alive. He thought that trees communicate and help each other, as they need each other[184]. Some of this research work is contested and needs further repeatable research and proof, it is important to add.

Which raises a tough question – 'How does the consciousness, sentience & intelligence profiles of the citizens of a city ruined by war, compare to the consciousness, sentience and intelligence of nearby forests?'

It is tempting to answer that the forests (with no brain as we conventionally know it) have a lot more consciousness, sentience & intelligence than the humans in the city ruined by war (filled with humans with brains that we can physically identify).

Why not? Is it not time humans compared their outcomes to other creatures' outcomes? Outcomes matter.

Yet at the same time, as we are talking about intelligence, consciousness and sentience on different *profiles*; we are talking about different sets of attributes; and so you cannot compare the people of a city with the trees of a nearby forest in any meaningful way, or on any single scale.

The point of using the *profiles*, is to see intelligence, consciousness and sentience in their *context*, warts and all, and not as a single scale. So seeing a demolished city tells a story of humans' profiles as being one that is both intelligent but destructive. Whereas the forest has a different tale, from a different set of profiles, as the trees are organisms that can only communicate and react slowly; and use that power to help each other.

The absence of a single intelligence, consciousness and sentience scale does not allow us to escape the shame of human violence and destructiveness; but gives us a chance to see other creatures as having different intelligence, consciousness and sentience, with more nuance & complexity than we understood it to be beforehand.

The breakthrough in thinking here is that Sentience, Intelligence & Consciousness *profiles* are a new and better way of *rethinking* these things.

The idea of using profiles to better understand these complex ideas still sits on the sideline of current thinking, and there is a strong story about researchers starting to use this approach as the *key* and the starting point towards reaching a true-to-life *perspective* on these matters.

Venn Diagrams Can Play a Role

To add another dimension here, we can look at Venn diagrams. John Veen was a mathematician who popularized these diagrams in the 1880's, as a good way of showing the logical relationship between sets of data, and these diagrams came to be named after him. If you can imagine a Venn diagram, with 3 circles, as below:

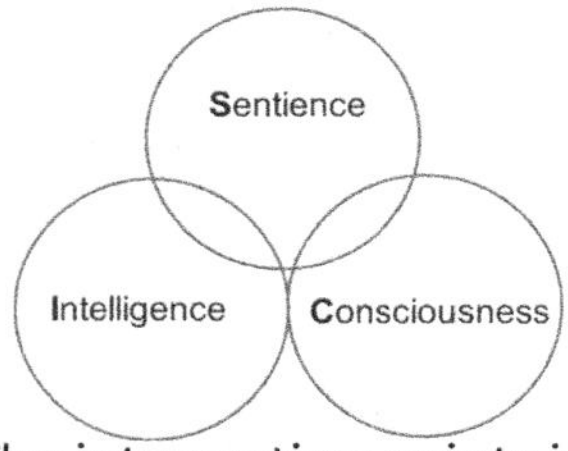

It's a diagram showing the possible logical relationships between sentience, intelligence and consciousness. (Note – this is not ruling out that other elements could be added to this diagram, but focusing on the 3 elements here does sharpen and clarify the discussion).

The intersecting points in the Venn diagram show where there is an overlap between the 3 elements. These intersecting points are usually represented by the ∩ character. This is a Venn diagram symbol. That is not an 'n' as in 'n' for Nancy.

One part can be <u>*written*</u> as I ∩ S, to describe the point where a creature sits in relation to its intelligence and sentience profiles – where they intersect.

That part of the diagram can be <u>*said*</u> as 'Intelligence intersection Sentience', where the ∩ = 'intersection'. (ie. equation **I** ∩ **S** translates to I=intelligence ∩ = intersection S=sentience). It's like a subset of the data, for example where a creature has characteristics that are clear evidence of both sentience and intelligence.

Similarly, where there is unity between the sets of data, and we want to view and understand all items in the sets, whether sentience, intelligence or consciousness; this can be represented by the character ∪. That is not a 'u' as in 'u' for umbrella. There are lots of other ways of using a Venn diagram too.

In the content of *profiles*, the above Venn diagram is going to be different for every living thing, based on each creatures' different intelligence, sentience and consciousness profiles.

And the Venn diagrams concept tells us about another *dimension* – that the degree of overlap and degree of unity in these different sets of skills and natural abilities that define sentience, intelligence and consciousness – adds another level of *complexity* in understanding the degree of each, as overlapping skills and abilities can make it hard to assess the level of intelligence and consciousness and sentience that form part of the overall profile.

It's a much more nuanced understanding of living things. All sorts of variations are possible – for instance a living thing may not need to be intelligent to be conscious, and you may not need to be conscious, to be intelligent. It's all buried in the profiles, different scales for sentience, intelligence and consciousness.

It's a different way of looking at things. Seeing other living things through the mirror of sentience, intelligence, & consciousness profiles, is a real way of *understanding and reimagining* what other creatures are really like.

And acknowledging that their space in the world is not ours to give or take, by right. As in the end, the SINGLE imaginary scale of sentience, intelligence, and consciousness, is gone, and can now be discarded as meaningless.

In particular, the low recognition and value of *sentience* is the most shocking of all this review. For example, that *pain* inflicted by humans unintentionally or intentionally, is felt by many creatures, but still discounted as not 'really' pain, as we can't attribute 'human' emotions to other creatures. Really? Is that an argument?

And where a creature is not only sentient but has some *intelligence* and *consciousness*, how does that change the experience for the creature? It could magnify the emotions, including the suffering and drama. The wild deer caught in a barbed wire fence, or the elephant who steps on a landmine, are both in pain, but sentience could turn this experience into unquestionable and deep awareness of their predicament and heightened fear.

We need to be able as humans to see other creatures in all their depth & complexity. And many creatures do start to look more complex and sentient and conscious and intelligent once the *single scale* is put in the dustbin of history.

But will we ever leave that SINGLE imaginary scale of sentience, intelligence and consciousness behind us? Hopefully we will! There is a need for more subtle understanding, to see the dark & light but also the shade. To seek more knowledge & awareness.

Starting to think about profiles of sentience, intelligence and consciousness, and how they *intersect*, for each living thing, will help us understand more about life.

Profiles let us dig down, getting the whole picture of each characteristic of living things, with all their nuances, and peel away layers of complexity.

It's a *vanishing* of the old *'single scale'* view of sentience, intelligence and consciousness.

6. Rethinking Biological Life

Biological Software

'*Biological life*' features (as per Chapter 1 – Rethinking Living Things) can be defined as an aggregation of different molecules that has a boundary isolating a special internal environment, that can reproduce itself, that consumes nutrients, stores energy for later use, releases waste materials, can sense and react to external conditions, grows new structural material, and dies.[185]

But as we have examined (as per Chapter 2 - Rethinking DNA); living things are not just chemical combinations of DNA; they are essentially made up of biological instruction sets.

To describe DNA programming, the best description is not *'digital biology'*, it is really & fundamentally '*biological software*'. There's a difference. Putting *software* at the heart of the equation means that everything makes better sense – it's all programming of one form or another.

Recognizing 'being alive' as a function of 'biological software' is a new idea. As 'being alive' has arguably been the most important biological phenomenon in earth's history, so seeing the rise of living things as a story of *biological software* evolution and what impacts it, is quite a mindset change.

It leads us to start the rethinking process, during 'Chapter 3 - Rethinking Evolution' and 'Chapter 4 - Rethinking the Brain,' in the context of *biological software*, and how we understand it now.

So instead of the original general terms like *'biology'*, or *'DNA'*, or 'genes' that all pre-date the full emergence of modern computers, software and the internet world; we now have an opportunity to re-describe what is happening in terms of *biological software*.

And already we are starting to use different software tools to build and manipulate things - using *biological* software tools on biological software; just like we now use *digital* software tools on computer software.

There still is a difference between gene editing on living cells and writing software on a computer; as one is living (gene editing) and one is not (computer software based in computers).

But how long will that clear distinction last? It's a convergence, a *vanishing*.[186]

Intelligence, Sentience & Consciousness

Despite our emerging understanding of the importance of biological software in how biological life works, there is so much we don't understand.

The human brain is still a mystery, despite many good theories being proposed and a generation of research.

To get more insight into the behaviour of living things, we need to understand and rethink intelligence, sentience and consciousness.

Understanding how these attributes work in living things today may give us some insights into how they could work in future non-biological life forms.

Studies of intelligence, sentience and consciousness have been held in a 'straight jacket' of a *single scale* for each, with humans at the top of this imaginary scale, and everything lined up below us, which was not making much sense.

The *vanishing* of that scale frees us up to rethink these key attributes.

The use of profiles lets us rethink each characteristic, whether intelligence, sentience or consciousness, and start to see what they really are.

As we move on to study non-biological intelligence sentience and consciousness, (in the next section), we can use the foundations of the above study of the biological world to help us.

— Part B —

Rethinking Machines & Modern Software

7. Rethinking Machines

A Modern Definition of Machines

A machine is defined as a piece of equipment with several moving parts that uses power to do a particular type of work.[187]

It's a solid but still vague starting point to understanding machines. That definition excludes what we would call tools (like a hammer or a scythe or an axe) that although powered by humans, have no moving parts. On the other hand, it would include something like a toaster, which has various moving parts used as the bread is toasted; and it is powered by electricity.

Clearly more complex machines like ovens, cars, trains, engines, heaters, and of course computers, all easily fit into the basic definition of a machine.

No mention is made in the basic 'machine' definition of the notion of autonomy; the ability to make your own decisions without being controlled by anyone else.[188] Another good term for autonomy is 'self-governing'.

This omission of any reference to autonomy when discussing machines was more related to the times that machines first appeared, from around 10,000 years ago when farming began; up to recently, where it was taken for granted and not even considered, that machines could one day be independent and/or autonomous.

In recent times autonomous machines have arrived, most notably in the form of driverless cars. They are autonomous in that many typical driving decisions have been handed to the computer system. Car sensors 'see' what surrounds the car, and navigate along the road, keep to the speed limit, keep to the lanes, & navigate traffic lights, all based on a '3D model' of the car and its surroundings.

These cars are autonomous but not independent, in that they still fall under the driver's control. It is just that many navigational decisions are passed to the vehicle's computer software system. So the system driving is not only automated, it can make decisions.

It will take a while for definitions in this area of driverless cars to be clarified, to allow for driverless cars with different capabilities. Changes to the law are going to be playing catchup for some time. A car that is driverless but needs to be watched over or monitored by an individual is going to be treated differently to a fully self-driving vehicle, or a vehicle with a lot of automation but still assumes a driver presence.

So the idea of machine autonomy and/or independence is still very new. A machine does not need autonomy or independence to be a machine. But more machines – especially computer-based machines, are starting to have some form of autonomy built-in to them. For example, self-monitoring software that can run diagnostics and sort out some issues itself, or the drone about to operate autonomously within very wide guidelines, and of course the driverless car.

Machines, especially computer-based machines, are growing smarter. And part of that story is intelligent machines having some autonomy. So autonomy would be a good addition to the definition of an intelligent machine. That doesn't mean total autonomy, which could be defined as full independence.

The terms independent and autonomous were developed over time to refer to people and countries – not machines. And currently, as Pediaa[189] explains it, the main difference between autonomous and independent is that autonomous refers to self-governance, while independent refers to not being dependent on anyone.[190]

You can have a machine that is self-governing (autonomous), like a driverless car, but not fully independent, in that you may allow for a driver to have a role in offering guidance. But it is still defined as intelligent as the car can make purposeful responses to available information.

Fully independent or fully autonomous machines will come later, and will most likely need to be called more than intelligent. A machine that is independent is starting to look like another life form. More on that later.

The Emergence of Machines

First there came the so-called *agricultural revolution* and the rise in agricultural productivity from new crops and new methods.

This began from around 10,000 years ago, when hunters & gatherers began domesticating wild plants, and became farmers in the middle east; which led to farming of new crops and new farming methods slowly spreading across the rest of world, and a world full of agriculture.

Instead of spending all day hunting for animals, many now had time to spend on other things, leading to the rise of cities, more complex hierarchies, ideas of wealth, and even organised religions and political ideologies. Hunters & gatherers became fringe dwellers in the modern world, in the Amazon, or forests of Papua New Guinea.

Then from 1940's to the 1960's came even further increases of agricultural production, based on high-yield varieties or cereals, as well as greater use of fertilizers, pesticides, and more irrigation, supporting larger and larger populations.

But while new agricultural machines also added some value, (eg. the Dutch-improved Chinese plough that needed fewer horses to pull them), there was no revolution in technology for agriculture, just occasional innovations.

It was really the *industrial revolution* across Europe & the USA from the mid-1700's to mid-1800's that changed the world and moved most hand-done production to new machines.

These were machines generating steam power, producing iron, manufacturing chemicals and textiles, all technology innovations. And the first machine tools and mechanized factory systems appeared, the precursors of the Ford Model T car production lines during 1908 to 1927, and the modern automated production lines for cars and all sorts of day-to-day consumer products used today.

The industrial revolution led to increased trade and a massive income growth of ordinary people, creating a higher standard of living across Europe, the USA, and many other places in the globe.

In historical terms, the *industrial revolution* is a massive milestone, just as important as the agricultural revolution.

A model of the spinning jenny machines that made spinning cloth faster by reducing the amount of work as workers could spin multiple spools at the same time'. It allowed cotton manufacturing to done in factories with multiple machines and workers, with greatly increased production. By Markus Schweiß, CC BY-SA 3.0,

https://commons.wikimedia.org/w/index.php?curid=195501

Almost everything in our modern lives comes out of industrial revolution – in the cities the food we eat passes through an industrial process, with our transport, houses, & modern appliances coming out of complex industrial processes or production lines. Millions of blue collar and while collar jobs work in the process of building or making products, to selling or managing their distribution and delivery across the globe.

It truly was a revolution that gave many the ability to afford to consume more; and live in more comfort. Whether these are 'better' lives than those of the hunter gatherers is a fascinating question, one not really asked until recently.

In any case, prior to the rise of computers from the 1940's; industrial machines had a strong history and contribution to the workplace and economic growth.

Computers didn't arise out of nothing; they were the next steps in the emergence of machines.

From the Abacus to Babbage

People have been trying to make calculations easier for thousands of years. The *abacus* was developed from earlier mechanisms used in Babylonia around 2,400 BC, to help with arithmetic. The *slide rule* emerged around 1620, invented by an English minister named William Oughtred, also to help with division & arithmetic. Like the *abacus*, the *slide rule* progressed over time, speeding up calculations and transactions for generations.

But it was Charles Babbage who first thought of the idea of a programmable computer. It came out of his work on what he called a 'difference engine', built for navigational calculations in the early 1800's.

The original computer was essentially a mechanical calculator, receiving a program and data via punch cards, and with a printer for output that could plot curves. What he called his 'analytical engine' had features like an arithmetic logic unit, and integrated memory.

It has been described by many as a century before its time. But with thousands of parts that needed to be hand-made, it all took too much time to build and in the end lost funding for further development His son completed a simplified version in 1906 but too little too late.

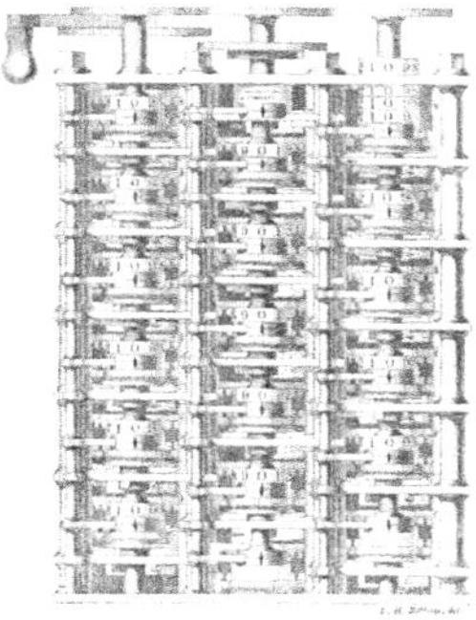

Portion of Mr Babbage's Difference Engine No.1. Source:

https://books.google.co.uk books?id=Fa1JAAAAMAAJ&pg=PP8&im ..#v=onepage&q&f=false

From Z2 to Colossus to ENIAC

Digital computers arrived in the late 1930's, first powered by electric switches driving mechanical relays (electromechanical), then all electric powered, using vacuum tubes. The Z2 computer of German Konrad Zuse, from 1939, is an early example. His 1941 Z3 version was programmable. His 1950 Z4 version, delayed by World War II, used the digital (0,1) number system and was sold for commercial use, making it the world's first commercial computer.[191]

During World War II the British code breakers broke the German *Enigma* and the *Lorenz SZ 40/42* encryption machines, the key to German military communications, and most accept that the information coming from this may have even changed the course of the war, or if not, at least shortened it by several years.

To solve the ciphers (algorithms used for encryption or decryption) of the *Enigma machine*, (see picture below) and so read German military communications, electro-mechanical *bombes* were developed.

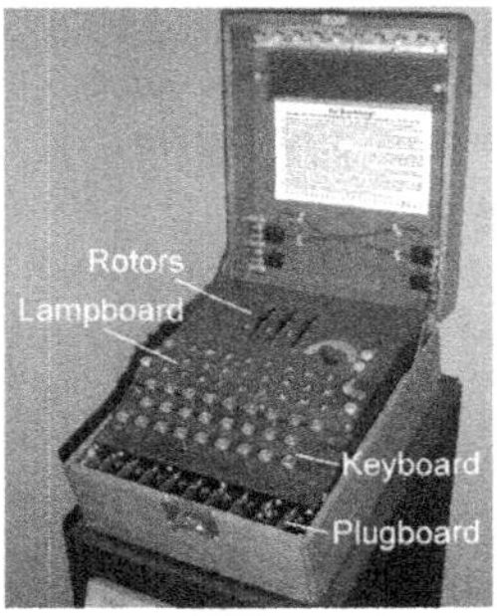

Enigma Machine at the Imperial War Museum, London.
Public Domain

The *bombe* was an electro-mechanical device originally developed by Polish code breaker Marian Rejewski before the start of World War II, then Alan Turing (more on him later) from Bletchley Park came up with a British design that ended up being an enhanced version as the team improved it.

A Bombe at Bletchley Park during WWII

The British version of the *bombe* was operational from March 1940 and was instrumental in breaking many German military codes. It achieved this by helping find the daily settings of the Enigma encryption machines, as set by the German armed forces.

In computer terms, the most important computing machine was Colossus, often described as the first programmable digital electronic machine. See below in 1943 at the code-breaking operation in the UK at Bletchley Park:

A Colossus Mark 2 codebreaking computer being operated by Dorothy Du Boisson (left) and Elsie Booker (right), 1943. This picture was created by the United Kingdom Government and is in the public domain.

British engineer Tommy Flowers was commissioned to build Colossus in 1943, to tackle the more advanced ciphers from German military command, that used the *Lorenz SZ 40/42*. It was operational at Bletchley Park on 5th February 1944. There were 10 versions in the end, each faster than the last. It used a large number of vacuum tubes to assist with logic applications, and was much faster than the *bombes*.

Even though this was developed later in the war, it was still significant in adding a big advantage to the Allies in deciphering German military code in the last stages of the war.

ENIAC was like a USA version of *Colossus*, but independently developed. Building was directed by John Mauchly and J. Presper Eckert at the University of Pennsylvania, around the same time as Colossus, though *ENIAC* was fully operational a bit later in 1945. It had a few important differences from *Colossus* - it had more flexible design, and much greater speed.

ENIAC was used to perform ballistics trajectory calculations for the US Army, in WWII.

Both *Colossus* and *ENIAC* could use a 'program'; actually, a process of manually setting switches and plugs. The switches had to be reset (re-programmed) to solve another problem. Not like the modern program based on a stored electronic instruction set.

But these were the first digital electronic machines for calculating, brought into being by the military demands for code-breaking in World War II.

Out of WWII

So just before and during World War II, in the 1930's and first half of the 1940's, modern computers were *separately* invented in 3 locations at around the same time. The Z2 in Germany, the Colossus in the UK, and the ENIAC in the USA.

Even without the war, the computer revolution was coming. But the war was the key acceleration factor that changed everything, and poured dollars into the further development, especially in the USA and UK, (much like the AI Arms Race today).

In Germany, that had lost the war, computing industry expansion was delayed, as the country rebuilt out of the ruins, and the Z4 came out in 1950.

In the UK, the Official Secrets Act stopped some key participants from trading on their reputations and using all their expertise to boost the computer industry there.[192]

And, of course, Alan Turing died in 1954, aged only 41, from a possible overdose or accidental poisoning. This took away from the world one of the leading thinkers and inventors in the computer scene. Nobody knows what additional momentum would have been created had he lived. A lot!

But in the USA, which had played a key part in winning the war, and came out of the war relatively unscathed, the urgency to develop computers for the war effort supercharged the industry; and kick-started the computer industry after the war. So the USA became the 1st home of the modern computing industry.

World War II and its aftermath gave the USA a head-start in the world of modern computers, that it is yet to relinquish.

Recent Chinese efforts in the artificial intelligence (AI) area are the first major challenge to that dominance.

From Computers to the Internet

The modern computers that emerged out of these early machines would end up taking a fraction of the space and be faster than imaginable at the time.

They all would come to have:

- a central processing unit (CPU), based on a microprocessor, to process information and instructions,
- memory storage, based on semiconductor chips, to store information,
- an operating system that talks to the hardware and provides services for the software programs
- a software program that allows the computer to perform a generic range of tasks/operations
- peripheral devices like a screen, mouse, printer, keyboards, camera to allow input and output.

One of the main milestones of the rise of computers was the *stored software program*. Having a stored instruction set might seem obvious now; but wasn't at the time. As mentioned before, using early computers involved rewiring the machine, if you wanted it to do something different.

Alan Turing came up with theoretical solution to that problem in 1936, in his brilliant paper 'On Computable Numbers'.[193] He proposed the 'universal computing machine', now called the 'universal Turing machine'. Its key design feature was a *program* (instructions) stored in memory (tape at the time).

If the instructions (stored program) changed, then so did the outcome. It could calculate anything. It was programmable. The idea of software became the heart of the modern computer.

One of the other main milestones in the rise of modern computers was the invention of the *integrated circuit chip* by Robert Noyce at Fairchild Semiconductor in 1959. These chips built larger and larger numbers of transistors into a small single silicon chip. Today the latest of these can fit billions of MOS (or metal-oxide-silicon) transistors on your fingernail, in a single chip. These chips became cheaper to build, and storage capacity and speed kept growing, based on Moore's law (the 1965 observation of the co-founder of Fairfield, Gordon Moore, who noted that the number of transistors on an integrated circuit was roughly doubling every 2 years).

The use of miniaturized microprocessors took computers everywhere. Computers started appearing on everybody's work desk (1970's), then in everybody's homes, (from the 1970's and 1980's) and eventually in everybody's hands as laptops, (from the 1980's beginning with the Osborne 1 in 1981), in a truly remarkable and rapid journey that nobody predicted.Beyond that came the 'network of networks', the invention of the internet. Its beginnings were in the packet switching APRANET, a 'backbone' for academic and military networks in the 1970's, which grew and extended into decentralised fashion into business and our day to day lives. Concepts like Internet Protocol address (IP address) and Domain Name System (DMS) helped regulate the growth.

A key stepping-stone in the rise of the internet happened when Tim Berners-Lee invented the World Wide Web (WWW) in 1989. It led to the world wide web on the internet became becoming publicly available on 6th August 1991.[194]

The original browser Mosaic, released in 1993 by Marc Andreessen and Eric Bina, kick-started a whole communications revolution.

Then new web sites were created using Hypertext, and the early browsers like Mosaic allowed people to find information quickly, transforming this small roadway into the full *information highway* we know today. Its reach spread wider and wider, from corporations, to governments and especially to the population at large.

By April 2022, there were 5.03 billion internet users worldwide, which is about 63% of the world's population. That 5.03 billion internet users included 4.7 billion social media users.[195] It keeps growing.

Technology moved down into everyone's hands – whether the PC, Notebook, phone, watch, or tablet, via the internet (and world wide web). The individual emerged as the centre of computing.

It was like the *'democratization'* of technology, where most individuals in the world have access to one form or another of modern technological tools; and this opened up millions of lines of communication across the world. Lots of room for personal data to travel around in. And lots of opportunities for people to capture and store all that information.

The *hardware* is the internet, and the software is the world wide web, but increasingly that line has blurred in popular culture, and many people tend to call the combination of the hardware and software the *'internet'*, even though it is not strictly correct to mix up the 2 elements.

We can all see the slow death of the newspaper, paper books, & eventually the fixed line telephone. We have seen the birth of online music, digital news, streaming videos, social networks, online shopping and more.

As households and businesses all placed computers at their hearts, the mobile cell phone, originally developed in the 1980's, also spread everywhere from the early 2000's. The original 'house brick' mobile cell phones became sleek fashion accessories and something most people can't live without, even while walking down the street.

Martin Cooper, inventor of the cell phone with his 1972 mobile phone prototype (as self-photographed in 2007)

https://upload.wikimedia.org/wikipedia/commons/1/1f/2007Computex_e21Forum-MartinCooper.jpg

The Internet of Things

Then followed the spread of computing power to all sorts of other devices, from fridges to vending machines, to QR codes, to car parks, and even shopping trolleys – the rise of the so-called '*Internet of Things*' (IOT).

It is getting bigger and bigger, now including wireless sensor networks, such as home automation, (from buttons to open the garage door, to shut the blinds or turn on the lights and heater), to building automation, that controls all the heating, lighting, and security in office towers.

Then there are commodity sensors on goods in stores, and smart speakers, wearable technology, remote Health monitoring for the sick and elderly and touch payment on trains, in shops or via toll sensors on the freeway.

In factories there are automatic process controls to improve productivity, and growing use of robots. On farms, data is automatically collected on rainfall, temperature, wind speed and more. In some harbours all ships are connected & automatically monitored. Smart industrial heating or cooling systems try to reduce energy usage. Imaging that can see through walls has also been released.

Drones have become the carriers of everything, from pizzas to bombs. They are the new 'eyes' in the sky for surveillance. They are the new weapon for war, for spotting and disarming tanks, or sinking ships. War by remote control is coming. Drones reduce privacy while at the same time add visibility to many long-range operations, like searching for missing people, or tanks.

The other side of this, is the rise of surveillance cameras everywhere – helping to solve crimes in the way the introduction of fingerprinting (as a form of identification) did from 1894 onwards, while at the same time strengthening the foundation of the surveillance state; and reducing privacy in new ways.

Machines are increasingly everywhere and increasingly connected. It is something we tend to take for granted, as it has been happening all around us for a generation.

The Digital Transition

In summary, from World War II onwards, to today, we have had what can be now called the *digital transition*.

This describes the *transition* from analog equipment to machines based on digital computing and digital communications. Digital logic, a series of 0's and 1's, became the standard for processing logic, powered by widespread mass production and popular use of microprocessors, modern computers, cellular phones and especially the internet.

People have called this 'digital transition' the 3rd Industrial Revolution, comparing it to the 1st Industrial Revolution (introducing machines to manufacturing in the 1700s & 1800's), and 2nd Industrial Revolution (introducing the railroad, telegraph & electricity in the late 1800's and early 1900's).

Some also see the arrival of new technologies like artificial intelligence, robotics, the internet of things (IOT), wireless technology, quantum computing and biotechnology as representing even an 4th Industrial Revolution, based on these emerging technologies.

This period since World War II was a time of transformation of society as well as how we work. Our lives and ways of communicating with each other, have changed forever, along with the changes to business and jobs.

Era's end?

Separating out the *3rd and 4th industrial revolutions* is quite hard to do however, as you need to find *defining* events and dates to separate them. You could easily pick 5 or 10 possible mid-point *'markers'* between 1945 and today; but calling them 'eras' just doesn't seem right – they are too close together in time, and a lot of change was more like continuous fast progress than earth-shattering changes. We don't really need to slice it up into too many 'eras'.

Seeing World War II as an effective *starting point* for an 'era' is logical - as the war was the impetus for faster computation and the need for urgent decoding, that directly led to the rise of the modern computer and created a pathway to today's computer-centric future.

The problem is finding a good candidate for the *ending point* for the era. Could it be 2023, when the global COVID-19 pandemic was finishing, after causing global turmoil and massive jobs disruption?

It doesn't sound right to pick a finishing year, to neatly tie off a period of around 90 years – from the start of World War II (1936 with the Japanese invasion of China, or 1939, with the German invasion of Poland) to the ending of the World COVID-19 Pandemic (2023).

It didn't feel like the *end* of an era.

That's because there are so many technological & social changes set into motion, and those changes are still moving fast.

What we have seen is the classic search for peace, prosperity and normality after the global conflicts of the 1940's, (now wearing off), and the wonder of the technological explosion of computer technology after that.

But the *digital transition* is still rolling on, getting even more momentum and speed during the pandemic, that accelerated the rise of remote access & remote working, while the internet continued to boom, as does the growth of the 'internet of things' (IOT). This technological era shows no sign of ending.

Machines Everywhere

The changes aren't over; innovation in the world of machines & computers are still moving very fast.

The significance of the emergence of more and more connected machines everywhere, (the Internet of things - IOT) is still being understood. The significance of smarter and smarter machines in our lives every day is still not fully understood. And the significance of the arrival of very intelligent machines, likely to be the most impactful, is definitely not understood yet, as it is just beginning.

All these machines are not something that we can easily phase out or turn off; as we rely on them more and more.

Is the human relationship with machines symbiotic [196]- having an *interdependent* relationship? Or alternatively is it a dependent relationship, where humans are increasingly dependent on machines?

It seems very much to be a *dependent* relationship, where humans have a growing reliance on machines. Machines do rely on us to keep going, for maintenance. But that may be a passing thing as more machines become self-maintaining. Where will that all lead?

We work with machines; we travel in machines, we carry machines with us, we live surrounded by machines, machines like TVs entertain us, we communicate by machines like phones and the internet, we use machines all day, unless we live in a forest or the countryside or by the sea and are self-sustaining and off the grid. Our lives are becoming more and more *machine centric.*

We are approaching the 'point of no return' in our relationship with machines. It's like a '*vanishing point*', where 2 lines converge, where humans can't live without machines, where humans have lost a lot of their independence in thinking, acting, working and living, and rely on machines for most things. A trend likely to accelerate as intelligent machines start to enter our lives more and more. Driverless cars are just the latest.

Machines are the 'wildcard', or unknown variable, for what happens next for Homo *Sapiens* and human civilization, as the *digital transition* rolls over us, and machines take on more and more roles.

Dependency can be good but can also be deadly. Ever-present machines and emerging more intelligent machines are causing a *vanishing* of human *independence* as we once knew it before the presence of machines, as they became irreplaceable in our lives.
Aside from the issue of growing human dependence on machines, there is another new fact. AI software is already intelligent. The new AI emerging on the latest computer servers might have the capacity to sustain consciousness and even life. Not bionic consciousness & life, but machine consciousness & life.

This makes AI potentially very different from earlier inventions like the spinning jenny or the ENIAC computer.

Soon we might have the machines housing the AI that are the foundation of a new class of conscious beings. They will make an impact of some kind, the possibilities of which will be explored in the coming chapters of this book.

As that happens, there will begin a process of the *vanishing* of human dominance of the planet. How Homo sapiens handle this, will shape the future.

8. Rethinking the IT Superpowers

Software became a *part* of many people's lives in the last decades of the 20th century. From then it has risen to take up a *central* role in almost everyone's lives.

Below is the background story of the pioneers who started it all, and the entrepreneurs who built it into what it is today.

A/ THE PIONEERS

Software had many faces as it *emerged* to take its place in the world.

Below are some notes on a few pioneers – including a few giants - who appeared, and helped transform the industry.

Ada Lovelace

Software was being written even in the 1800's, but not in the modern form that we now know it.

Ada Lovelace is credited with being the very first programmer. She was the daughter of famous poet Lord Byron and Annabella Bryon. She worked with Charles Babbage from 1843 on an advanced calculating machine called the Analytical Engine.

She wrote what we would call today the first computer program, to calculate Bernoulli (calculus) numbers, and imagined it could be used for other things.

Watercolour portrait of Ada Lovelace around 1840, Artist Alfed Chalon (likely)

Science Museum Group, Creative Commons Attribution 4.0 licence (cc)

She was definitely the exception to the rule, as computing from then onwards tended to be dominated by men, and even today female developers only form a small percentage of total programmers at any time.

But her memory is worth reviving, to help inspire young woman into a computing career. She has shown the way.

Alan Turing

Alan Turing was another pioneer at the dawn of computing software. He developed the core theory of modern computing in the 1930's, which proposed that *software* was going to have a separate life to the *hardware* and *operating system* of computers, a concept at the heart of modern theoretical computer science. He conceived the idea of a *'general purpose'* computer, effectively a computer that with different programming, could do different things.

Turing was an English mathematician, who became a cryptanalyst then computer scientist. He was born in 1912, raised in southern England, and left Cambridge University with a degree in Mathematics, then PhD in Mathematics at Princeton University in 1938.

He became famous for his role in helping break the German naval codes in World War II at Bletchley Park. As part of that role improved on original Polish tools (so-called bombes) to find the settings of the German Enigma machine faster, so the code breaking turned into practical intelligence that helped win the war, in key encounters like the Battle of the Atlantic between supply ships with their destroyer escorts versus German submarines.

In 1948 (after the war) he worked on designs for ACE (automatic computing engine), a design that was based on stored programs, then later he worked on the Manchester computers, that ended up building the world's first stored-program computer.

In 1950 he wrote an article called 'Computing Machinery and Intelligence', ('Mind', October 1950[197]). In it he proposed an experiment now known as the 'Turing Test'. It was the first effort to see if a *standard* could be set to define machine 'intelligence', (as a way of answering the bigger question of whether a machine can think).

His idea was that if a person could not tell, through conversation, that they were 'talking' to a human being or a computer, then the computer had passed the test of being *'intelligent'*. He saw it as a version of the *'imitation game'*, (which usually asked someone to choose, only from text, and sitting in a separate room, whether they were 'talking' to a man or a woman). 'The Imitation Game' later became the name of a movie (2014) about his life.[198]

His 1950 paper was pioneering, as Turing started thinking about defining and measuring machine intelligence; and developed early ideas on artificial intelligence (AI). We come back to this topic later in this book.

In 1952 he fell victim to British laws that prosecuted individuals for homosexual acts, leading to archaic hormone treatment on him as punishment. He died supposedly of suicide in 1954 of cyanide poisoning. No-one is really sure what happened.

So tragically he was never fully recognised in his own lifetime, as his achievements in World War II were covered by the Official Secrets Act, and his new work was cut short by his premature death. What could he have discovered back in the 1950's and 1960's, had he lived, particularly in the area of artificial intelligence (AI)? We will never know.

Alan Turing at Princeton University in 1936. Anonymous Unknown author, Public domain, via Wikimedia Commons.

https://upload.wikimedia.org/wikipedia/commons/1/17/Alan_Turing_%281912-1954%29_in_1936_at_Princeton_University.jpg

His premature death meant that he was not able to continue working on new computers, new software ideas and new design concepts, which ended up being a massive blow to the growth of the computing industry in the UK, and partly due to this, the global leadership in computers and software passed to the USA. UK attitudes to homosexuality at the time ended up costing them their own Silicon Valley.

Thomas J Watson (IBM)

IBM started out with electric tabulating machines (that summarized information stored on punch cards) and time clocks, but really starting to grow rapidly under the leadership of Thomas J Watson Sr, in the period 1914 – 1956. He knew how to manage, and how to train and manage salespeople. Some of his ideas came from his former boss John Henry Patterson from NCR.

A big milestone in the company's success story was when they built the IBM 701 in 1952, their first commercial store-program computer based on vacuum tubes. Then came machines with a hard disk drive (IBM305 from 1956) and use of transistors (7000 series from 1958).

More interestingly from the software side, IBM developed the FORTRAN (standing for 'formula translating system') programming language for general purpose applications, but good at scientific and mathematical calculations especially. It became the foundation of many other languages like BASIC.

John Backus was the IBM employee who in 1953 proposed to management that this FORTRAN language be developed, to make programming easier than assembly language programming, (which operated at a lower level, often one statement per machine instruction and was complex to write, use, and was specific to each computer).

FORTRAN was the first of a generation of 'higher level' software development tools, that let programmers write code without worrying too much on how the computer actually did things in explicit detail; and could also work on different computers.

IBM began supplying both smaller and larger corporations of the USA with computers to speed up calculations and boost productivity, especially with their IBM System/360 from 1964. They were the first big company to build new computers that still could run the existing software, the upgrade concept that was later deployed by many hardware and software vendors.

IBM became the dominating supplier of large computers from the 1960's to the 1980's, when the PC and generation of smaller & less expensive computers emerged. Over that time they invented the ATM, hard disks, floppy disks, Dynamic Random Access Memory (DRAM), and of course the SQL programming language and the relational database. An amazing mix of hardware and software innovation.

And they made progress on artificial intelligence (AI) with IBM Watson, one of the first conversational chatbots, that could answer questions asked in natural language, which famously won the quiz show Jeopardy in 2011.

IBM is still a massive corporation today, after going through many changes, including selling its personal computer business to Levono in 2005, and its x86 server division to Levono in 2014.

IBM's main business today is computer services and software, and their 'glory days' of being the most dominant computer and software company are gone. But they remain a force, and their legacy is totally underestimated today.

They will be forever remembered for Thomas Watson's brilliant sales approach to getting new business and keeping existing business. The salesmen were famously dark-suited & well-presented, and trained to have enormous pride in, and loyalty to, the company, and pay particular attention to good service.

Thomas Watson Sr circa 1920, from the IBM Corporate Archives.

Licenced under the Creative Commons Attribution – Share Alike 3.0 Unported licence. Attribution IBM

QS:P,+1920-00-00T00:00:00Z/9,P1480,Q5727902s

Watson's favourite slogan was 'THINK'. It came from a dispirited discussion at sales meeting in 1911, when he worked managing sales at National Cash Register (NCR), before he left for IBM. He said sales staff don't think enough.[199]

At that 1911 meeting he spun around at the front of the meeting and wrote 'T-H-I-N-K' with a blue crayon on the easel behind him. He later said that by 'THINK' he meant that he wanted salespeople to take everything into consideration[200].

His boss John Henry Patterson saw the easel the next morning, and told Watson to put the sign in every Department in the company before noon the next day.

Patterson was a boss who counselled 'big ideas and small words', and Watson's THINK slogan was exactly the sort of thing he loved. Nice having a boss with a big vision.

IBM Chairman Thomas Watson is also famous for something he did not say - that there was a world market for 5 computers. Multiple sites refer to it as coming from a meeting in 1943, but it most likely came from a misquote at an IBM shareholder's meeting in 1953. Watson actually said that as a result of his trip, he expected to get orders for 5 machines, but came home with 18 orders.

That became, over time, according to the great website 'Geek History', the infamous remark and urban myth - that there is a world market for five computers.[201]

Tim Berners-Lee

Born in 1955, Tim was brought up in an English household with his parents being talented mathematicians, who both worked with early computers, nurturing his early interest in the burgeoning area. He left Oxford University with a degree in physics but built computers in his spare time.

He then got a job at CERN (the European Organization for Nuclear Research), that operated a large particle physics laboratory. While there he invented the world wide web (WWW), the current global standard for accessing information over the Internet.[202]

Along with friend Robert Cailliau, they developed a mark-up language called HTML (Hypertext Markup Language), for creating web pages and web applications, as well as the URL (known as an address on the web).

He also conceived HTTP (Hypertext Transfer Protocol), the protocol allowing computers to exchange information with each other[203].

On 6th August 1991, Berners-lee first posted, on Usernet, an invitation to the public to collaborate with him on the WorldWideWeb project. The rest, as they say, is history. The world wide web spread out to cover the planet and be used by billions of people as the way they interact with the Internet.

Tim Berners-Lee arriving at Guildhall ro receive the Honorary Freedom of the City of London

https://creativecommons.org/licenses/by-sa/4.0/

Tim Bernes-Lee went on to be a prominent advocate for more individual control of data, and an improved web he called the 'Semantic Web' that proposed to extend the web to include document *metadata* like dates of articles and when they were written and who wrote it, which are hard to search for now. AI may in the future bring some of this greater flexibility in finding data to life.

He won the 2016 Turing Award, 'for inventing the World Wide Web, the first web browser and the fundamental protocols and algorithms allowing the Web to scale'.[204]

He played a unique role in the birth of the modern web, in that he did not try to turn his invention of the world wide web into money for himself, and develop a patented & licenced offering, but instead let it become available to all, a decision from which millions benefited.

He married his second wife Rosemary Leith in 2014. They are both founding directors of the World Wide Web Foundation. They support AI companies and also collaborate on venture capital.

B/ THE ENTREPRENEURS

Following the pioneers came the business people, the ideas people, who were able to build amazing businesses based on the new infrastructure and technology.

Bill Gates (Microsoft - Ms Office, Azure hosting, LinkedIn. OpenAI)

With the 1960's a new wave of entrepreneurs arrived, along with a new wave of technology.

1968 was the time of Hippies & the Vietnam War and was the 1st time an eighth grader called Bill Gates got to do real-time programming at Lakeside, a private school in Seattle for elite families there. He was born in 1955 to a father who was a wealthy lawyer, and a mother who was the daughter of a banker. The Mothers' Club there decided to start a computer club, and got the state of the art.

It was very unusual to have real-time or live access to a computer at the time. Most programmed via entry of computer punch cards, but time-sharing systems were invented in 1965, allowing 100's to access a single system by sharing the total time available – effectively live.

Having the ASR-33, linked to a mainframe in central Seattle, meant that Bill Gates did real time programming in year 8 in 1968, and spent all his time there. Then in 1971, when he started college, he and his friends gained access to an ISI (Information Sciences Inc) computer for unlimited access, as long as they helped write some payroll software. In one 7 month period in 1971, Gates and friends ran up 1,575 hours of computer time on this ISI mainframe, which averages out to 8 hrs per day, 7 days a week[205]. Bill Gates was obsessed, skipping athletics and staying up all night.

So it was good luck that Bill Gates had hours and hours to program, at the early stage of the modern computing revolution, but he also had the aptitude to make it happen, spending long hours honing his craft.

Malcolm Gladwell, whose stunning book of 2008, 'Outliers – The Story of Success',[206] rewrote the story of Bill Gates origins. This book was a best seller of the time, and offered an explanation of why and how some people rose to amazing success.

He compares the story of Bill Gates with the story of the Beatles. Again, very talented, but through circumstances, given extraordinary opportunities. The Beatles were invited to Hamburg in 1960; and went 5 times between 1960 and 1962.

On the first trip, they played 106 nights, five hours or more each night. On the second and third trip they likewise were on stage for hours, and had to work past their top songs, into playing just about everything from cover versions, as well as rock and roll. They ended playing a total of 270 nights on stage in 1 ½ years[207] A lot of long sessions playing together.

Malcolm Gladwell called this the 'Hamburg crucible'[208], that set the Beatles apart. They came back from Hamburg with more stamina, confidence, and discipline on stage. They sounded amazing and original - then burst onto the music scene in 1964 with successive hit singles.

Similarly, Bill Gates set up in his business called Microsoft with Paul Allen, after thousands of hours of programming, and dropping out of college. Malcolm Gladwell called this phenomenon the 10,000 hour rule,[209] where heavy and timely exposure to their craft, set them up to be leaders in their fields.

In 1975, Popular Electronics magazine announced the launch of the Altair 8800 for $397, a DIY computer. The modern computer age had begun. Bill Gates at aged 20 was ready to take advantage of it, and with childhood friend Paul Allen founded Microsoft in 1975.

Bill Gates in 2017. By Kuhlmann /MSC - https://securityconference.org/en/medialibrary/asset/bill-gates-1523-18-02-2017/, CC BY 3.0 de, https://commons.wikimedia.org/w/index.php?curid=100184908

But to go from boy genius the be one of the richest men in the world, takes a lot more than a great starting point. For a start, he had his grandmother's motto 'Think smart!', very reminiscent of Tom Watson's THINK motto at IBM. But also he had a legendary reputation for hard work. (Later Apple would use the slogan 'Think Different' from 1997-2002 as a counter or answer to the IBM slogan).

He helped make the software revolution take-off. It started with the development in 1975 of the BASIC programming language for the Altair 8800, (see picture of the early PC below), which was developed only after a demonstration to show it was arranged, starting a pattern of bold marketing, that sometimes meant products were announced for delivery by a date which was then delayed, then delayed again.

Late delivery of promises became a hallmark of early Microsoft, deployed with devastating effects on competitors who thought the dates of product releases were more than just press releases.

The Altair 8800 computer (on the left of the above picture) plus the Model 33 ASR Teletype as terminal, paper tape reader, and paper tape punch. Image courtesy of Boston-based collector Timothy Colegrove. For more information about this image send an email to timothy.colegrove@gmail.com

Nevertheless, Microsoft's Altair BASIC was a hit with the many computer hobbyists who emerged at the dawn of the modern computer industry.

IBM approached Microsoft in July 1980 regarding software for the new personal computer about to be releasee, the IBM PC. They wanted Microsoft to write a BASIC interpreter for the new PC, and later to provide an operating system, that ended up being called PC DOS.

Critically, and cleverly, Bill Gates did not offer to transfer the copyright of the PC DOS operating to IBM, and fatally, IBM did not insist on it – after all, they paid $50,000 for it so they should have negotiated some kind of ownership.

So when other hardware vendors produced the IBM-compatible PCs that swamped the market, MS-DOS, the version of PC DOS Microsoft sold for use on IBM compatible PCs, became the industry standard and this led to Microsoft taking off.

The *Windows* operating system followed in 1985, aiming to compete against Apple's Macintosh graphical user interface (GUI).

The original Windows release was so delayed that it popularized the term *'vapourware'* - software that exists mainly in the mind of developers rather than being real.[210] But it was such a success story in the end, now used by billions across the planet.

Today, after years of successful growth, Microsoft is still known for its Windows operating system, but now also for it Microsoft 365 suite of office tools like Word, Excel, Powerpoint etc, and its Internet Explorer & Edge browsers for internet use.

Bill Gates transitioned out of Microsoft management between 2006-08, to spend his time on philanthropy and other projects. As of 2021, his net worth was $152 billion, according to the Bloomberg Billionaires Index. He & wife Melinda divorced that year, after 27 years of marriage, so that wealth may end up being split in two, as there was no prenup[211].

In 2022, based on revenue, Microsoft became the world's largest software maker, (but not the largest computer company, which was Apple in 2023 with $US 385.09 billion). In the 12 months to 31st March 2023 Microsoft revenue was $US 207.59 billion.[212]

Steve Jobs (Apple - iTunes, Apple music, hardware, TV)

The story of Steve jobs role in Apple's rise to become one of the largest companies in the world is a legend in itself.

Steve Jobs was part entrepreneur and industrial designer, with some wonderful sales and marketing skills. But underneath was a person of great intuition and ability to innovate, with some famous flaws. (For some of his time at Apple it was a bad idea to get into the lift with him. Some staff found themselves grilled and then fired by the time the lift got to the ground floor).

He was a 'baby boomer', born in 1955s, (the same year as Tim Berners-Lee and Bill Gates), a child of the protest movement and flower-power generation, who flirted with Eastern spiritualism in the form of Zen Buddhism in his youth. He embraced it to an extent.

He was a pioneer of the personal computer revolution with Steve Wozniak, and especially in the success of the Apple II. He was the force behind the Macintosh, the iPod, the iPad & the iPhone. Each one of those machines featured some stylish design and innovation.

But not everything he touched turned to gold, and he did have some time in the wilderness from 1985 after sailing Apple close to the rocks and losing an internal power struggle. But he bounced back brilliantly in 1997, and returned to Apple and delivered its 'golden' era of product innovation, fast growth and great profits.

Steve Jobs on-stage at Macworld conference, San Francisco 2005. This file has an extracted image: Stevejobs Macworld2005-cropped.jpg.This image was originally posted to Flickr by mylerdude at https://www.flickr.com/photos/mylerdude/3563553/. Licensed under the terms of the cc-by-2.0.

Steve Jobs is also known to some, amongst his other stellar accomplishments, as '*the man who stole the mouse and the GUI*' from Xerox.

This story is so buried in myth that it is hard to unravel it. Steve could see the potential of the Xerox Alto (their PC) in 1979, which had the first mouse and graphical user interface (GUI). Did he steal both ideas and take them back to Apple, as goes the story in popular imagination?

He definitely got to look at the Xerox Alto in action in 1979, after he had a 2 hour inspection of some of the Xerox innovations, (over the strong protests of their local manager)', and the original ideas seen there, were used in the failed Apple Lisa in 1983 then the much more successful Macintosh in 1984. Is it a clear-cut case of theft? No, not really.

There was more to it. Jobs struck a deal. He got few hours walking around the Xerox facility that ended up by looking at the Xerox Alto personal computer, complete with a demo of the working mouse and graphical user interface (GUI), *in return* for offering Xerox 100,000 shares in the upcoming IPO for Apple, for $1 million dollars.

The demonstration at Xerox moved the cursor around the screen with a mouse, moved Windows, clicked icons on the screen to open new functions, showed a modern word processor and even sent an early 'email' to a colleague in their world's first Ethernet network. It was the whole WIMP (windows, icon, mouse, pop-up) effect.

Of course, Steve Jobs took those brilliant ideas – the mouse and the GUI - back to Apple!

But it can't be called stealing, it was a commercial transaction – shares in Apple's IPO for time to explore. In the end it was a very dumb transaction from Xerox's point of view, as something they had been working on for years was suddenly in the hands of a competitor, without the heavy time and dollar R&D cost. Apple just had to innovate around the edges to make it work commercially.

However, it does need to be mentioned that at the IPO in 1980, 100,000 Apple shares were worth $22 each, so Xerox made money on day 1 of the IPO ($22 x 100,000 shares = $2,200,000, and they only paid $1m. So Xerox earned $1,200,000 profit on the day of the IPO). In addition, if they held onto those shares, which were trading in early 2024 at $195 per share, then their holdings would now be worth $19,500,000 ($195 x 100,000 = $US19,500,000, ignoring the impact of inflation). Still definitely a healthy profit by doing the deal.

Yet letting Steve Jobs see the mouse and GUI in action was not a great move if they really knew the potential of what they had, at the time, (for a 2 hour inspection), but that is only obvious in hindsight. Their own Xerox Alto PC using that mouse and GUI, bombed out when launched, as it was too slow and too expensive. But they might have had another go at it if they could see how important it was.

Compare that with the effort of Steve Jobs, who *innovated* straight away on what he had witnessed[213]. The original Xerox mouse cost $300, had 3 buttons and was easy to break. The Apple model cost $15 and had 1 button, and a lot longer life.

And the GUI? Well, the team at Apple improved that too, making it simpler to use by removing intermediate menus, and allowing users to grab the corner of a window and make it bigger. It became so easy to use, that once you saw that interface, you never wanted to go back to typing instructions with a cursor.

Author Malcolm Gladwell also picked up on a further point – *urgency*.[214] Steve Jobs was keen to get it market, as he could see it was so innovative.

So, in the end nothing was stolen. But the mouse and GUI were sold/traded very cheaply, because if Xerox kept the mouse & GUI away from Steve Jobs, they might have become the next Apple, one of the largest companies in the world. And Steve Jobs, a guy running a start-up business at the time, might have vanished into the pages of history.

The story of *'the stolen mouse and the GUI'* has become an urban legend; but doesn't take anything away from Steve Job's massive legacy.

Steven jobs died of a respiratory arrest in 2011, at the age of 56. His death is believed to be related to his pancreatic tumour, diagnosed in 2003, but left untreated for 9 months while he tried alternative natural cures. He left behind his wife and 3 children from that marriage, and a daughter, from an earlier relationship, whose name (Lisa) was given to one of the earlier Apple computers.

Jeff Bezos (Amazon - retail catalogue esp. books, Kindle and cloud computing)

Jeff Bezos was born in Albuquerque New Mexico in 1964, and educated at Princeton University. He became a successful manager at well-known hedge fund D.E. Shaw when he quit to try his hand at his own business. He was 30, his wife MacKenzie was 24. They packed up their lives and moved from New York to Seattle in 1994.[215]

He wanted to start an online bookstore, just as the internet was taking off. It began in his garage at the back of their house, with 2 desks he put together himself, made from $60 wooden doors purchased at a nearby Home Depot store.

He had some savings, plus $US 245,000 from the 'bank of mum and dad', an investment his parents Mike & Jackie made in the new business[216]. That's a lot of money in 1994, and tells of trust, belief and definitely devotion. Where would any of us be without that bank?!

They launched in 1995 and book orders started pouring in, sometimes increasing 40% or 50% some weeks. They rode the new internet, at the right place at the right time. Book sales in a new online world. Jeff Bezos concentrated on increasing cash flow and growing market share, around the idea of 'Get Big Fast'.[217]

Rapid growth led to an IPO in 1997, and his management style featured fast decision-making and enthusiastic adoption of new technology, as a way of meeting surging demand and getting the books out to buyers in a timely way.

They rode the 'dot com' boom then bust of the early 2000's, surviving low cash flow, but always growing while other 'dot com' startups vanished. A $100 million cash injection from online service AOL in 2021 took them to safety, then profitability in 2003[218].

Then some big decisions by Bezos started to transform the business and really propel growth:

- Amazon opened their web site to 3rd party sellers and list books and other products alongside Amazon goods, so attracting new shoppers while taking commission from these other businesses.
- Bezos hired experienced executive Jeff Wilke, and together they turned warehouses into 'fulfillment centers', allowing better prediction of customer orders, addition of more product categories like jewelry and clothing, as well as guarantee of faster delivery.[219]
- Bezos worked with key Amazon executive Andy Jassy, to build stable and infinitely extensible computer infrastructure, that ended up becoming Amazon Web Services (AWS), selling 'hosting' services to other companies who needed computer power. A bold and brilliant initiative that has boomed.
- Bezos worked with executive Steve Kessel to start its own digital book reader called the Kindle, which again turned into a great success.

Brad Stone, author of the brilliant books 'Amazon Unbound' and 'The Everything Store', regarded the confluence of the 3 main initiatives above - fulfillment centers, AWS and the Kindle – as the key to vaulting Amazon[220] into the ranks of the other big 4 (Apple, Facebook, Google, Microsoft).

Then consumer electronic devices like Kindle, Echo smart speaker with Alexa, Fire Tablets and Fire TVs and increased the reach of Amazon into many households.

Many other services were to follow, including Zoox (autonomous vehicles), Amazon Prime Video, Amazon Music, Amazon Publishing, Amazon Studios, then Blue Origin, a private space company. All innovative and growing businesses in their own right.

A core part of Amazon success was the way they tracked customer preferences, and then let them know of similar books or other products in that area. It was a new idea on the world wide web, and the process lent itself to being done automatically, and really supercharged book sales. Amazon laid the foundation for everyone to follow.

Customers who love vampire books are the ones most likely to be interested in new books in the same genre, and so finding the connection between buyer and new product makes for an easy extra sale.

By 2021 Amazon turnover of $469,822 million[221] passed that of Walmart, and by 2023 Amazon was the world's largest online retailer.

Jeff Bezos in 2017 from Van Ha. Public Domain.

https://www.dvidshub.net/image/3908584/jeff-bezos-visits-laafb-smc

Not everything was achieved without pain. 18,000 staff were fired in 2023 to cut costs. Criticism of Amazon has been made of working conditions in Amazon warehouses, some environment practices, and management efforts to stop unionization.

Criticism was also made of the impact of Amazon on physical bookshops, obviously affecting the viability of many smaller vendors. The effects of massive growth inside the company and its impacts outside the company, are still playing out around the world.

In 2019 he divorced MacKenzie Bezos and she received 25% of Bezos's Amazon shares, worth $36 billion. Jeff Bezos stood down as CEO in 2021 and became Executive Chair of the Board. Jeff Bezos net worth in 2023 was $US 150 billion, and he is the 3rd wealthiest person in the world after Elon Mush and Bernard Arnault. In 2023 he sold down his ownership of Amazon to 9.56%, around 25% of what he owned in 1998.

Amazon remains a truly amazing success story, and a tale of bold and imaginative management.

Sergey Brin & Larry Page (Alphabet – parent of Google, inc. eBay, YouTube, Google Docs, Google Maps)

Sergey Brin was born in 1973 and arrived in the USA, aged 6, when his family migrated from Russia. Both his father and grandfather studied mathematics, to which he added computer studies at the University of Maryland.

Larry Page was also born in 1973, in Michigan. His father was a computer science professor at Michigan State University, and his mother as a computer programming instructor there also. As a child he immersed himself in the world of books, magazines, and computers, with parents that fostered creativity and encouraged invention.

Both Sergey and Larry definitely had a great gene pool with a mathematical bent and inquisitive natures, a great starting point in life.

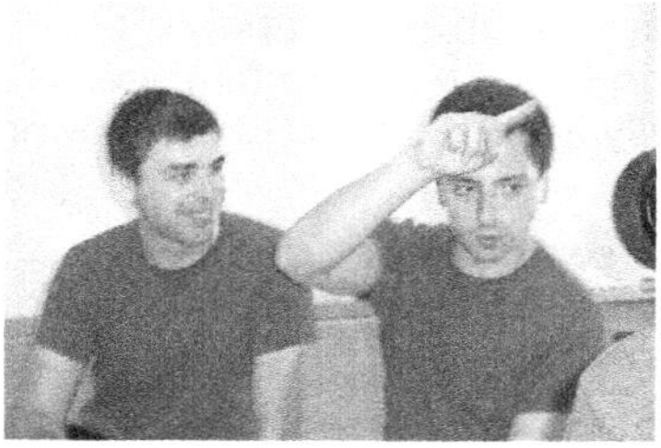

Larry Page and Sergey Brin in 2003 at around 30 years of age.

Ehud Kenan, originally posted to Flickr as 100_0334. CC BY 2.0 Generic.

<https://creativecommons.org/licenses/by/2.0>, via Wikimedia Commons https://commons.wikimedia.org/w/index.php?curid=3901443

They met as Stanford University PhD students in 1995, when Sergey was given the task of showing (new student) Larry around the campus. They got arguing & debating, and by 1996 they were partners, who suspended their studies and began building a search engine that ended up being named Google. It was based on links that determined the value of individual pages on the world wide web – the *PageRank* product.

The Google product was a hit at Stanford University and attracted a lot of investor interest as its use and success grew. Then Sun co-founder Andy Bechtolsheim put in $100,000 and Google Inc was founded in September 1998. They left the dormitory behind and moved into their first office in Menlo Park, then to Mountain View, California. They remained unorthodox, with an love of corporate bright colours and even had Yoshka, the company dog.

Google Search became the dominant search engine in world, because of the quality of its searches, based on its algorithms. People liked doing a query and getting back a list of web sites that were most relevant to them. The use of Google Ads allowed vendors to pay money to get their web site mentioned as *sponsored*, rather than let the 'organic' search algorithms put them, say, on page 50 of a person's query, where they would never be found. It became a gold mine, with Google being the most visited web site in the world alongside Facebook and Twitter/X.

They grew so fast that they listed on the US Stock Exchange in 2004, a super-fast rise to success. A share purchased for $85 in 2004 is now worth around $5,400. The Economic Times says that a $16,000 investment on Google in the 2004 IPO would now be worth $1 million dollars.[222]

Alphabet (Google's mother company) then expanded into other products, such as Gmail, web browser Chrome. and bought YouTube, and moved into artificial intelligence, cloud computing, cloud storage (Drive) and e-commerce (eBay), consumer electronics (Nest), wearable technology (Fitbit), navigation (Maps), language translation (Translate), and many other products. They now sit as one of the most valuable brands in the world.

Recently Page was estimated to have a net worth of $124 billion in 2024, while Brin has a net worth of $114 billion. Brin has been married and divorced twice, and has 3 children, while Page is married with 2 children. Mother company Alphabet had around 180,000 employees at the end of 2023.

Such a transformational company like Alphabet is very unusual. Google's search engine still dominates the internet, and their efforts in the AI area will be fascinating to watch. They have invested heavily in this zone and are bound to have a big impact.

Products like ChatGPT are changing the nature of searching, as people can get more detailed replies rather than a list of web pages that can help them. But Google will deal with that challenge with its own 'generative pre-trained transformers' (GPT) offering.

Mark Zuckerberg (Meta - Facebook, WhatsApp, Instagram)

Mark Zuckerberg was born in 1984 in New York to a psychiatrist mother and dentist father. Like Bill Gates, he started using computers and writing software in middle school. At Harvard in 2002 (which he attended briefly before dropping out and moving to California), he as seen as a computer prodigy, and launched Facebook in February 2004 with his roommates.

Mark Zuckerberg in 2005.

By Elaine Chan and Priscilla Chan - Derived from: File:MarkZuckerberg.jpg, CC BY 2.5, `https://commons.wikimedia.org/w/index.php?curid=27431613

It was originally designed for college campuses, but rapidly spread out far beyond that, to the point that by 2012, Facebook had reached 1 billion people and was listed in an IPO with Mark Zuckerberg as majority shareholder.

Meta (the Facebook mother company) has around 3.1 billion people using it daily world-wide[223]. Mark Zuckerberg is now married with 3 children, and his net worth in early 2024 was $167 billion.

Clouds on the horizon include the management of Facebook content (sexual & political), and the cost & prospects of the metaverse. Latest figures show it has cost Meta around $36 billion dollars to build the metaverse, (via its Reality Labs division) and it has not delivered much for that vast investment.[224]

Jyoti Mann of the *Business Insider* magazine points out that a source inside Apple say it spent $150 million to build the iphone,[225] a lot less. Though these comparisons are hard to verify. Nevertheless, the metaverse needs to show it is making headway and is the shape of the future soon, to keep justifying the scale of the expense. There is a case for arguing that until the avatars are more 'true-to-life' looking, that the metaverse will not kick off. In which case there is a long way to go.

Investment in achieving a great virtual reality (VR) experience, is another big Meta R&D expense on emerging technology, that needs to prove itself. Meta has a good offering in this area, the Meta (Oculus) Quest VR headset. The

Apple Vision Pro is another great product too. An idea sometimes proposed, is that these optical displays will eventually replace everyday computers for most people. That idea is sounding very optimistic at this stage of development.

So whether the investment in these areas is going to bring a massive return is still open for debate. But it is early days and these investments could deliver products that go mainstream, if a 'killer app' is found for them. Online gaming is an obvious candidate. Pornography is another obvious candidate, if the history of the internet is any guide. Making things more 'true-to-life' will be important here too, as it is for the Metaverse.

Marc Randolph and Reed Hastings (Netflix)

Marc Randolph was born in 1958 in New York, the son of a financial advisor father and real estate firm owner mother. A perfect background for someone who would become a leading entrepreneur. He graduated with a geology degree but in his first job in 1981 got into the direct mail marketing business, and using computer software to track customers' buying behaviour.

So very early on, (just like Jeff Bezos pioneered the book recommendation engine for Amazon), Marc Randolph could see the importance of capturing and using market research as part of the day-to-day user experience with Netflix. He also picked up on the idea of overnight delivery improving retention of customers. He more he learned, the more he wanted to repeat the success of Amazon, but not by selling books, by selling Compact Disks (CDs).

Reed Hastings, on the other hand, was born in Massachusetts in 1960 , the son of a lawyer father and mother from an elite Boston family. After college he joined the Peace Corp, and taught in Africa, then did a computer degree at Stanford. After that he was involved with various software companies either as founder (Pure Software in 1991) CEO or CTO and picked up a lot of managerial experience in the process.

Reed Hastings came up with the Netflix idea, and so Marc and Reed teamed up in 1997 to co-found Netflix, so bringing together Marc's journey into understanding the mail-order world, and Reed's IT and management expertise, often learned the hard way.

Marc Randolph co-founder of Netflix speaking in Arizona

https://commons.wikimedia.org/wiki/File:Marc_Randolph_by_Gage_Skidmore.jpg

The marketing and technology breakthrough was the focus on selling DVDs (not VHS cassettes and so easier to send in the mail) via a web page to collect orders and details of customer preferences (ie with no mail out catalogue), so combining 2 new technologies, and ending up with a great new business model.

The idea of Netflix as a flat $ rate for use per month (regardless of usage) movie rental by mail, run from a web page, was brilliant, and the right plan at the right time, with the right new technology. It all seems so obvious now, but that's what makes an entrepreneur great, they can see it before we can.

When Reed Hastings saw YouTube going well in 2007, Netflix launched their video streaming service to stream shows (moves, TV shows) to computers, which eventually led them into becoming the leading streaming service in the world today. Streaming became a 'game changer', Marc Randolph said in an interview. "We knew that giving people the ability to watch what they wanted, when they wanted, was the future of media consumption".[226]

Reed Hastings in Berlin 2015

re:publica from Germany, CC BY 2.0 Copyright: re:publica/Gregor Fischer

https://creativecommons.org/licenses/by/2.0>, via Wikimedia Commons

Netflix by 2019 had grown its streaming service to 139 million subscribers in over 190 nations.[227]

Marc Randolph calls himself 'an invisible face' when it comes to Netflix. But his contributions to the company's foundation and early growth were invaluable, especially as CEO to 1999.[228] He left the company in 2002, after working on the IPO to list in 2000. Marc is married with 3 children, and his current net worth is estimated at $100 million.

Reed Hastings is married with 2 children, and now lives in California. In 2023 he stepped down as Co-CEO of Netflix, and is giving away parts of $4.1 billion dollar fortune to worthy causes.

Netflix has become a story of how innovation and risk-taking with a new business model can transform an industry. Their innovation continues, as they move rapidly to make content and bring movies from places all around the world to our screens.

Robin Li and Eric Xu (Baidu)

Robin Li was born in 1968, and studied information management at Peking University, then Computer Science at the University of Buffalo.

He worked for IDD Information Services from 1994-1997, which was a leading internet search provider from 1996 – 2001. Part of his job there was to improve algorithms for search engines. In 1996, (at IDD) he created the RankDex. This was a pioneering algorithm that ranked or 'site-scored' web pages to allow better and more accurate ranking of them by the IDD search engine, to give better results for people searching.[229]

Few know that *RankDex* was cited in Larry Page's patent application for Google's PageRank as the first qualitative search engine. Larry Page was a Google founder, and he filed his patent for a search algorithm in 1998, 2 years after Li had created *Rankdex*. *RankDex* technology was licenced back to Robin Li for the Chinese search engine *Baidu*. (It was also licenced to Microsoft's Bing).[230]

Baidu was launched in China in 2000, and the 2 founders were Robin Li (search engine expert) and Eric Xu. Eric provided the funding and general support to the project, as a senior businessman and entrepreneur. It is interesting to note that Baidu was the first to give advertisers the opportunity to bid for sponsored Ad space, in return for paying Baidu every time a customer clicked on the Ad, so predating Google's advertising method.

Baidu took off fast, becoming widely used across the large Chinese domestic market, to the point where it became the biggest search engine in China with 80% of the Chinese search market, almost matching Google's dominance of the search engine query market in the rest of the world, which some put at 90%.

In 2007, Baidu was the first Chinese company listed in the NASDAQ-100 index, and in 2024 had a market capitalization of $36 billion.[231]

On the back of that success, in 2024, Robin Li, Chairman and CEO of Baidu, had his net worth estimated at $US6.2 billion by Forbes (His wife's net worth was counted separately at $1.7 billion). He is married with 4 children.

Robin Li at 2010 Web 2.0 Summit

This file is licensed under the Creative Commons Attribution 2.0 Generic licence. https://www.flickr.com/photos/kevinkrejci/5186169609/ This image, originally posted to Flickr.

Note – No picture available of Eric Xu Yong, Co-Founder, Baidu, that is copyright free

As Baidu grew even larger, big plans took shape, like development of a driverless car and buses, and development of a ChatGPT competitor in the Ernie bot. Ernie 4.0 chatbot came out in 2023.

Like all fast-growing companies, there were some setbacks on the way. In 2010 some Iranian (No-one is really sure who they were) hackers paralysed the search engine and took it off the air for hours. And some Baidu workers were arrested in 2012 on the claim that they were taking bribes to delete posts from a forum.

But mainly it has been a story of growth, and adding on additional products like Baidu Maps, Baidu Wangpan (cloud storage service), Baidu MP3 Search and Baidu Music (for songs), Baidu Translate, and Baidu Wallet (mobile wallet in partnership with PayPal).

Pony Ma (Tencent - WeChat)

Pony Ma was born in 1971 & brought up in Hainan Province in China, one of the poorest in China, by his father, who was a cadre and accountant for the Hainan Port Authority, and his mother, who became Tencent's legal representative, and was known as a strong woman.[232]

Pony Ma's real name is Ma Huateng, but as that surname translates to 'horse', he is known a Pony Ma, a kind of humorous translation.

He was one of the original 5 founders[233] of Tencent in 1998, and became its best-known person. Yet few would see him or know his face, or his influence.

Author Philip Renard felt that his low profile is by design, partly as he tries to keep the Chinese Government happy. But also, because he is, by nature, someone who does not seek the spotlight.[234]

Ma Huateng (known as Pony Ma) at Tech Crunch 2011 in Beijing

TechCrunch, CC BY 2.5 <https://creativecommons.org/licenses/by/2.5>, via Wikimedia Commons

After he left university, he was keen to come up with some business ideas to capitalize on the emerging internet, and found something called ICQ, an internet chat platform. ICQ was founded by an Israeli company called Mirabilis in 1996, then bought be AOL in 1998 and finally by Mail.Ru (VK now) in 2010. The chat service could be downloaded by anyone for free. It was the first mass instant messaging service on the internet, allowing a private conversation in private on-line, quite revolutionary at the time.

So in 1998 Pony and his 3 friends (co-founders) launched their own chat service for the Chinese market, basically copying ICQ, but calling it OICQ (adding an Q).

As author Philip Renard points out, it was common practice in China to clone Western services (Baidu copied aspects of Google, Alibaba copied aspects of eBay, and Weibo copied aspects of X (Twitter). So Tencent copying ICQ, simply stealing someone else's idea, remained a fundamental part of Tencent's strategy.[235]

However, when we think back to Steve Jobs use of the Xerox mouse and original Windows GUI, while he did not steal it, (he sold shares to get a look at it); but the original idea was only worth so much, it needed a lot of rethinking and refining to make it really useful and financially successful. The same applied to OICQ, they had to tailor it to the Chinese market.

So they made the download file much smaller, and thus faster for users to download. They allowed users, even if they changed computers, to hang onto their friends. They improved the search capability in looking for people[236]. They also allowed strangers to connect if they wanted to, an idea that proved to be a foundation stone of the yet-to-emerge social networks. In 2 months they had 500 users, then suddenly thousands were joining every day, and they were on their way. Tencent started to concentrate on OICQ.[237]

By 1999, they had reached 1 million users, but were not making money as servers were very expensive, and it was a free service. By their second year they had gone from 1 million to 50 million users, but were still not making money.

Then Pony Ma found a Korean web site called SayClub.com, which allowed personalized avatars, which could have their own customizable appearance and clothing, so they copied that and incorporated this into QQ. Chinese users could now have their own unique profile, to suit their interests and character. Extra designer clothes and accessories were charged to users at a very low price, but these tiny transactions were so popular with especially younger uses that they put them in profit.

The growing use of QQ as a dating app also meant that these accessories became part of that process too, where people wanted to show how they were different, or what their personality was like.[238]

Tencent went on to dominate the gaming market, by integrating games into QQ. However, many if not most people were sharing games for free, and so not paying for them. Tencent solved this issue by making the games free for everyone, and make money through 'in-game' sales instead – like features, upgrades, extra lives etc. Today these are called *microtransactions* and are commonplace in many games. But Tencent were the first to push them heavily.

They were engaged in a constant search for new apps and ideas that they could copy and add into QQ. American tech companies did the same thing, but as Philip Renard points out, China at the time was like the 'wild west', and anyone could copy anything from anybody and expect no protection, with Tencent earning the reputation of being a 'killer copycat'.[239]

The biggest app was still to come. With the growth of mobile phone use in China from around 2010, Pony Ma was worried that someone would come up with a mobile based chat app, that would put the existing desktop based QQ out of business. So they decided to develop *WeChat*.

Pony explained later at a Conference, that ideas are not important in China – execution is. You had to get your idea to market fast before competitors could match you.[240] So they added things like 'Moments', their version of Facebook for personal posts, images, stories, articles & media feeds. Then they added 'Friends Nearby', an early version of Tinder.

Later came the Red Packet function, allowing people to digitally send red packets to people, which was real money going into their bank accounts. When 75 million red packets were sent via WeChat one year, Tencent realised that mobile payments were going to be the key to future payments and launched WeChat Pay in 2013[241], their mobile payments platform for all digital payments.

WeChat grew to 100 million users in 2012, after less than 12 months in business,[242] a record at the time only exceeded by ChatGPT later.

In 2017, they added a Mini Programs feature, a copy of the App Store feature that other mobile providers like Apple have, except it is built into the WeChat app. It allowed users to access thousands of apps (or mini programs) inside *WeChat's* own app, for example a restaurant app or Didi, the main Chinese Uber.

WeChat became what Philip Renard called the *'everything app'*.[243] So you could go from playing games to travelling to dinner, booking a restaurant, catch the news, updating friends with pictures, then pay the bill, all via WeChat[244].

Will such an *everything app* emerge in the West?

It seems unlikely. Elon Musk was thinking about it, but with concerns about the power of big tech, and issues with social media, it is doubtful if anyone wants a single company with such a reach into everyone's life.

That being said, it is still highly logical to have one single interface for everything you do, and this author predicts that it will happen quite soon – but at the bot level not the application level. More on this idea in Chapter 11.

WeChat does have about 100 million users outside of China, but concerns about surveillance and censorship have slowed growth in this area, and scared potential users away.

The Chinese Government does monitor and censor information sent by WeChat's apps, especially on sensitive topics like protests in Hong Kong, where messages can be even totally blocked, or Tiananmen Square, or Tibet.[245] This is the dark side of the internet revolution.

China also has a 'social credit scoring system', where all you do, where you go, what you buy, who you are associated with etc, ends up giving you a score on how trustworthy you are as a citizen. Philip Renard puts it brilliantly – a billion Chinese have sacrificed all aspects of their privacy and free speech for convenience.[246]

Pony Ma's wealth is sometimes calculated at $38 billion, and comes from his 9.7% share of Tencent, plus his property & art collection in Hong Kong. He is known for his very private lifestyle, and little is known of his family, apart from that he met his wife on the Tencent chat software, and has one daughter.

Jack Ma (Alibaba - Taobao)

Jack Ma was born in 1964 in Hangzhou (south-west of Shanghai). His surname 'Ma' is the same as Pony Ma, (of Tencent/WeChat fame). His mother was a factory worker, his father a photographer.[247] As a boy he developed a love of the English language, that helped lead him on a pathway to becoming a teacher. At Hangzhou Teachers College he met the woman who was to become his wife, and he graduated in 1988 with a Degree in English.

Not really an entrepreneurs story at this point. But times were changing in China, as Deng Xiaoping returned to power and his famous expression 'to get rich is glorious' captured the new mood, where the government was now encouraging people to be entrepreneurial. Jack tried his hand at a few businesses by the mid-1990's with limited success, but while working as an interpreter, he got a month-long trip to America and got a glimpse of the internet in its early days, as email started to take off.

He came back with the idea of building a Chinese version of the online yellow pages, and set up a new business named 'China Pages', and left his teaching life behind.[248] But the internet had few users and poor performance at the beginning, so it was a struggle, and working with the government was uncomfortable. Jack lost control of the China Pages business to his SOE (state owned enterprise) partner and got out.

However, the Chinese Government, concerned about the fall of the Soviet Union in 1990,[249] (and subsequent turmoil) due in part to the technology gap with the West, started rapidly expanding the telecommunications network for mobile phones and broadband connections across China. The internet became a viable business platform.

Jack Ma founded Alibaba in 1999, when China only had 2 million internet users. By 2000 there were 17 million Chinese online. He was aiming Alibaba at providing a business-to-business directory, an online marketplace where buyers and merchants can link up, to buy and sell to each other. However this business model in China was a struggle, and Alibaba was still not making a profit.

Jack decided to get into the ecommerce market, mimicking Amazon or ebay.[250]

However a big competitor to any new ecommerce business had just arrived. In 2001 ebay came to China, taking over a Chinese company called EachNet, that had 90% of China's ecommerce market. To succeed against eBay, Jack Ma had to defeat eBay in the field, a larger and better funded company with a global reach and global success.

In 2003, Alibaba launched its ecommerce consumer offering, *Taobao (hunting treasure* in Chinese) with 8 staff in the *Taobao* team. This was a consumer-to-consumer (C2C) business It grew by word of mouth, in contrast to eBays heavier use of advertising. It would take years before it turned a profit. But it was popular with 'mom-and-pop' store owners and grew fast.

In the end, eBay didn't listen to suggestions from its Chinese partners in EachNet, and the new advisors from the USA mostly didn't speak Chinese, meaning that had a big learning curve before they were useful. Also eBay had no customer telephone number, as per the USA, which was not liked in China. And while eBay followed a 'google' style clean format, the more cluttered Taobao web site had a more Chinese feel to it, with pop-ups and floating banners.

Most importantly, Taobao was free for buyers and sellers, while eBay had fees built into their standard model of business that nobody liked. This feature attracted the merchants, who in turn attracted the shoppers. The next disaster was when eBay decided to migrate their servers from China to the USA, which made eBay a lot slower, and subject to Chinese Government 'chokepoints' when it felt like it. Another big blow to eBay was when Alibaba released Alipay in 2003, allowing basic electronic payments[251], and Taobao kept on innovating, introducing text messages letting people know their payment had gone through.

By 2005, eBay's share of the market had gone from 90% to around 30%, and Taobao was up to 60% market share[252]. By 2009 Taobao's share of the market was 80%. Finally eBay effectively gave up the fight and left eBay in the hands of a local partner, where it faded out of sight.

Alibaba got a huge cash injection from Yahoo of $1 billion in 2005, giving it the cash it needed to keep funding Taobao, at that time still not making a profit, as well as reward employees who helped make the Taobao success story happen.

Three (3) successful startup businesses – alibaba.com, Taobao and Alipay were all booming, to the point that a Microsoft offer in 2008 to buy the whole Alibaba business for $44.6 billion was rejected, annoying Yahoo enormously.

In that same year, Tmall was created (2008). Tmall was originally named *Taobao Mall* by Taobao in 2008, as a business-to-business (B2B) platform within the Taobao consumer-to-consumer (C2C) website. In 2010 Tmall then moved to its own website tmall.com, to distinguish it sellers, larger merchants and brand owners or their distributors, from Taobao's smaller consumer focus. In 2011 Tmall became a fully independent business, tmail.com, and in 2014 Tmall Global was aimed at international business.

The main focus then became Tmall (B2C), Taobao Marketplace (C2C) and eTao (shopping search engine). More recent times have seen Alibaba acquire media, and get involved with start-ups in venture capital roles.

Alibaba listed on the New York Stock Exchange in 2014, raising $25 billion, giving the company a value of $US 231 billion, which by 2018 had passed the $US 500 billion valuation, the 2nd largest after Tencent[253].

It is an amazing success story. Jack and his wife have one son. His net worth is reported at $30.1 billion in 2024.

Elon Musk (X, formerly Twitter)

Elon Musk was born in 1971 to a South African family. His father was a businessman but also had a lot of ups-and-downs, a mix of legitimate business and hucksterism.[254]

As a boy he was fearless and very competitive, but also occasionally moody. In the 7 years Elon lived with his father as a boy, up to age 17, life was difficult, as his father would many times call him a moron and idiot, and seemed to lack compassio[255]. Elon wanted to escape, so he migrated to Canada at age 18 (his mother was Canadian making it easy to get there).

But like any childhood, there were good times. As a young person he was an enthusiastic reader and at age 10, was a keen player of video games and computer programmer. At age 12 he sold his first game, written in BASIC, to a computer magazine for $500. He was a computer prodigy, but at the same time, he worked hard at it. When he hounded his father to buy him a VIC-20, an early computer, he got a 60 hour course in how to program BASIC (language). He completed the course in 3 days, barely sleeping.[256]

He went to university in both Canada and the USA, at the University of Pennsylvania, to gain a degree in economics and physics. He went to Stanford University in California in 1995 but quit after 2 days to start a software company and write an online city guide, as the internet revolution took off.

Elon Musk (cropped) in Colorado in 2022 CC

https://www.dvidshub.net/image/7132411/usafa-hosts-elon-musk

That Zip2 startup was purchased by Compaq for $307m in 1999, so starting Elon on the journey. He then founded X.com, a direct bank, which in turn was merged with another company to form PayPal, which eBay acquired for $1.5 billion in 2002, giving Elon another $100 million in the kitty.

He went on to found SpaceX, proposing to run services to and from space, and also helped create a solar energy company that was acquired by Tesla in 2016. He was also involved in founding OpenAI, an artificial intelligence research company, and Neuralink, a company exploring computer to brain interfaces. Finally he purchased Twitter in 2022 for $44 billion, rebranding it as X, and in March 2024, xAI was created, to further his interest in AI.

X (formerly Twitter) started in 2006 as a social media site for posting and sharing texts, videos, and pictures that are called 'tweets'. Its point of difference is that users who tweet have to keep things brief – sometimes called *microblogging*. Originally users could only put 140 characters in a tweet. But then in 2016 it allowed attachments and photos to be added on top, then in 2017 the maximum number of characters doubled from 140 to 280 characters, and in 2023 Twitter Blue users (users who pay a small subscription fee for extra services) could do a 4,000 post. But the concept of a small blog is still at the heart of a 'tweet'.

Elon Musk's acquisition in 2022 meant that X/Twitter went private, and began a series of changes including staff reductions, introducing new features, promoting free speech, and trying to reduce 'spamming' (the sending of unwanted messages to multiple people). His original vision for X (as outlined in 2022 while he was in the process of buying Twitter), was to combine a social network with a payments platform, and it seems likely that he will continue with that vision.[257]

Twitter growth was exponential, but did slow. And at special times, it peaked. For instance, when Michael Jackson died in June 2009, 100,000 updates per hour were being posted, that crashed the servers.[258] Today X is estimated as having 500 million users.

In summary, Elon Musk's rise has been marked by some controversy but has been brilliant in anyone's language. He has 10 children by four partners, including 2 wives. His net worth was estimated at $182 billion in 2021. That figure moves up and down a lot based on the value of Tesla shares.

Zhang Yiming & Liang Rubo (TikTok)

In 2016 a Chinese startup company called ByteDance released TikTok, [259] known as *Douyin* in China.

The people who founded ByteDance was Zhang Yiming and Liang Rubo. Zhang became the ByteDance Chairman, while Liang became CEO. Zhang was born in 1983 in Longyan, Fujian province on the south-eastern coast of China. He was the only son of civil servants, and ended up graduating in engineering at Nankai university in Tianjin in 2005.

He started work as an engineer for travel web site Kuxun, and after a few other jobs and a startup experience, returned there to run the Kuxun search business, then founded ByteDance with Liang Rubo in 2012. He had a vision of building a business that pushed content to users that they actually want, via AI-based recommendations. They initially built a news aggregator app called Toutiao.

Zhang Yiming, speaks at a charity event in Chengdu city, southwest China in 2015

Photo by ChinaImages. Covered under depositphotos.com extended licence

The along came the TikTok app in September 2016, which millennials (born from the 1980s to early 2000s) loved. The acquisition by ByteDance of Musical.ly, a popular social network service that had 100 million subscribers, helped turbo charge TikTok along, as all Musical.ly users were moved to TikTok.

Musical.ly let users build videos along with background music or audio samples. Users could record then share small videos of themselves lip syncing to popular songs, which they could also speed up or slow down with some simple software editing.

ByteDance acquired this for $1 billion.[260] ByteDance were able to do this as they became financially backed by some large financial groups, including Chinese investors, with a majority of the company (around 60%) in the hands of global investors.

TikTok (with Musical.lp features incorporated) took off fast, as the idea of a sharing network for *'quick music content'*, or short form videos, was a gap in the market (probably left by YouTube) that especially teenagers were looking for and wanted. Some have argued that it is popular because people don't have the time to watch longer videos. You can watch multiple videos within 10 minutes with TikTok, for example. Others suggest that young people have shorter attention spans.[261]

The TikTok format is user centric and easy to use. It allows a big range of content, whether lip-syncing or dance challenges or something educational. That diversity attracts advertisers.[262] Like with most apps, you create a *profile,* and start saving TikTok videos you like, or making them.

The type of content on TikTok is wide-ranging, part of its attraction, and new videos appear every second. Videos range from showing various 'challenges' (eg the watermelon mustard challenge, which consists of eating a slice of watermelon that is spread with mustard), to content produced using the 'I'll never forget you' trend, (to talk about someone who had a good impact on their life), to playing games (like Grand Theft Auto: San Andreas), to Legos hobby videos, to fashion, to fishing, to sewing, to relaxing walks, to showing collections (eg. of toys), fitness including health food, nail art, student content, techniques of building in video games like The Sims or Planet Zoo, house cleaning, personal grooming, cooking tips, pets and so on The list is endless, and so is the potential for people to upload things they find most interesting. Videos can be followed, liked or shared.[263]

The analogy with X (Twitter) is interesting, TikTok has short videos, X has short tweets. Many people like to make and also receive quicker content.

TikTok is all about user-generated content, and this has fuelled TikTok's popularity. Users make their own videos and then share them with their friends, so driving activity on the site, and more downloads from others. The other part of the success story are the TikTok algorithms.

They monitor a wide range of behaviour during video watching, to build insights into the users, and then present what they will really like back to them. For instance the 'react' feature lets users film their response to a video.

So they go further than (other algorithms in) spotting likes, follows and clicks, reaching another level of understanding of users. The intuitive layout makes it easy to keep going. In short TikTok are following a path pioneered by earlier very successful algorithms like those used by Amazon and Netflix, but have built new algorithms that add more insights into video users behaviour.

TikTok has become so successful that by 2021 the global community of TikTok users was around 1.1 billion users. It became the most popular website in the world in 2021 too, surpassing Google.[264]

TikTok's rise has meant questions from US regulators, with President Trump threatening to ban the app in 2020 over security issues – in 2019 they did get a $5.7 million dollar fine for collecting personal information on children under 13. This issue of Chinese ownership will probably come up again in the years ahead, depending on US – China relations. Zhang was different to many other Chinese entrepreneurs in that he focussed on global growth.

Zhang met wife at university, and has personal wealth of $44.5 billion, based on his 24% share of the company.[265] It is believed that the couple do not have any children. ByteDance is now valued at $250 billion, making it the most valuable privately held company in the world, according to Bloomberg[266]. Zhang stood down as CEO in May 2021. The company is still growing fast.

C/ THE ACRONYMS

Technology discussions used to be all about FANG (Facebook, Amazon, Netflix & Google) as the fastest growing US technology companies.

After 2017 it became FAANG, with the addition of Apple, another very fast-moving tech company. All these companies' share price roughly doubled between 2018 and 2023.

That FAANG acronym ignored all the Chinese internet success stories, who became known as BAT (Baidu, Alibaba & Tencent). But TikTok owner ByteDance is another new important player, so making BAT become BBAT, or BATT.

However, in recent years, another 2 companies stood out for growth and success, and people turned their minds to including them – *Microsoft*, which was excluded due to the lack of exponential growth before ChatGPT, and X *(Twitter)* that has a great growth story too.

Below is an attempt at an acronym to cover them all – using the term 'Batman'. When the Batman theme song is sung, they extend the word, to become BBBBAAATTTMMMAAANNNNN.

Thus, the idea of putting the whole 11 leading companies together:

BATTTMMAAANX – Baidu, Alibaba, Tencent, TikTok, Microsoft, Meta (Facebook), Apple, Amazon, Alphabet (Google), Netflix, and X (Twitter) sitting at the end.

So, in common use, the term could be *BatmanX*, but be actually spelt as BATTMMAAANX. A work in progress. But the 11 are all defined in one acronym.

These eleven (11) companies have emerged to dominate the world wide web:

a. Baidu (Chinese search engine)
b. Alibaba (Chinese online shopping – Tmall, Taobao)
c. Tencent (WeChat)
d. (Tik Tok) ByteDance
e. Meta (inc. Facebook, Instagram, WhatsApp)
f. Microsoft (inc. Azure Hosting, LinkedIn, ChatGPT)
g. Alphabet (parent of Google, inc. eBay, YouTube)
h. Amazon (retail catalogue esp. books, Kindle and cloud computing)
i. Apple (hardware of all kinds, iTunes, Apple music, big range of software)
j. Netflix (streaming service)
k. X (formerly Twitter)

9. Rethinking Web Software

a/ THE *ENGINES* DRIVING THE NEW WEB SOFTWARE

Like in any revolution, the rise and rise of world wide web software is fuelled by special individuals, circumstances in time, and 'engines' of growth.

Personal Skills & Right time, place and attitude

The founders of all the 11 companies above were at the right place, and the right time, with the right attitudes and skills, and found the right technology. All these fascinating people and businesses had a glimpse of the future dangled in front of them, and grabbed hold of it:

- Bill Gates of Microsoft had access to online computing when very few did,
- Steve Jobs of Apple saw the potential of the mouse and new GUI
- Pony Ma of Tencent found ICQ, that early instant chat tool, then used it as his starting point to build WeChat.
- Jack Ma of Alibaba saw the early internet and its potential on a visit the USA in the 1990's and came back with the idea of building an online 'yellow pages' business in China.
- Robin Li of Baidu got to work on improving algorithms for search engines from 1996-2001, in the early days of that technology, and that gave him a big start on most competitors.
- Reed Hastings at Netflix saw YouTube booming in 2007, and realised it was perfect to deploy on a streaming service.
- Mark Zuckerberg of Meta was, like Bill Gates, a computer prodigy, coding from middle school onwards, who also had driving ambition to succeed, found a niche, and quickly launched Facebook with his roommates in February 2004.
- Sergey Brin & Larry Page of Google dropped out and began building their own search engine, just as the internet was taking off, web sites were proliferating, and end users found sites hard to find.
- Jeff Bezos of Amazon started an online bookstore, just as the internet was taking off.
- Elon Musk was also a coding prodigy, and came up with some big ideas about direct payments, space research, then later acquired Twitter (X)
- Zhang Yiming saw the potential of the short format video and jumped on it to create TikTok.

All the above men were talented and hard-working. But mostly they were opportunists who saw a chance and took it. And were at the right place, at the right time, and with the right attitude & experience to grab hold of the opportunity with both hands. And in the case of a few, being a coding prodigy gave them a head start.

Engines of Growth

It is the *engines* of growth that are even more fascinating. Yes, the internet had arrived, and was changing the way we communicate with each other, and the rise of the mobile phone accelerated the idea of computers available 24 x7.

But beyond that, there were 4 critical engines of growth, easy to see now but very hard to spot before they arrived:

- Algorithm engines – With web pages multiplying by the thousand every day, how do you find them? Search tools based on algorithms. *Algorithms* are a sequence of instructions that solve specific problems, used especially on the internet to determine how content is filtered, or what people really like. Only a few realised how important they would be – and still are. Why waste time searching through piles of web pages when an algorithm can give you the best results on a few pages? Algorithm engines became the heart of Google, Amazon, Baidu & TikTok.
- Video engines – Being able to run videos on the internet was pioneered by companies like YouTube. Streaming technology started off very slowly but then reached the point of being viable as a business too. Apple's itunes, while launched in 2001, started supporting purchasing & downloading of digital music in 2003, and then in 2005 supported digital music videos, building a great video engine to help sell songs. (In 2019[267] itunes was replaced with Apple music on the macOS, then later for Windows users[268]). In addition, Reed Hastings at Netflix saw YouTube video success (2007) with streaming, then switched to sending his movies out on a streaming service. Likewise, Zhang Yiming at TikTok picked up on the demand for short format video and ran with it.
- Contacts engines – The idea that people wanted what became known as a *social networking service* was something new when Facebook came out on February 2004. Leveraging off the internet, products like Facebook, WeChat, X(Twitter), WhatsApp, Instagram & LinkedIn etc. created *contact engines*. This is software allowing people to share information, pictures, videos, stories and contacts, in any way, shape or form people liked, within reason. Never before had it been so easy to share this information, about friends or families or events. It became known as *social media*, and succeeded firstly because it allowed users to make their own content, and share it with whoever they wanted, and secondly, because it was interactive. People could reply or follow up, so it was much more immediate than an email, or (snail mail) letter.
- Online Shopping Engines – The ability to shop 'online' on the internet, became possible once people trusted the payment tools on the world wide web, and once more and more products appeared on the world wide web for sale. Payment tools include products like PayPal. Stripe, Square, FIS Global, Adyen, 2Checkout, Braintree, Checkout.com, Sage Pay & Amazon Pay. PayPal is still the largest, operating in over 200 countries and supporting more than 25 currencies. Products for sale on the internet

started appearing from the 1990's, to the point that now most retailers offer online purchases. Early developers like Amazon, eBay and Taobao (Alibaba) carved out important niches. The big success of newer players like *Temu* and *Shein* highlight the rapid growth going on in the sector. They continue to take business off 'store front' retailers. Online sales have been estimated at around 19% of total sales globally, but around 15%[269] in the USA, and up as high as 27%[270] in China. Eventually many people started ordering things like groceries and alcohol online, that previously had low online rates of order. The global pandemic from 2020 - 2023 sped up this process.

These 4 engine rooms of growth – algorithms, videos, contacts and online Shopping - sit at the heart of why the top 11 fast growing IT corporation are thriving.

b/ THE SHORT-TERM EFFECTS OF WEB SOFTWARE

The fast take-up of the world wide web on the internet made it one of the most ground-breaking changes in the way humans communicate since the widespread introduction of fax machines (in the 1980's), and before that, the printing press (around the year 1440).

There were some immediate effects:

i. THE YELLOW PAGES - The so-called *'Yellow Pages'* of countries moved online, with key vendors moving to offer all these services on their web pages. This covered things like businesses offering jobs, wanting to sell houses, wanting to sell cars, or personals (like dating sites). Previously these businesses were extensive users of advertisements in newspapers. The migration of Ads from the world of paper to the online world drastically affected the viability and business model of newspapers, leading to mass sackings.
   The eventual solution for newspapers was to also move online, and start charging subscription fees to those who wanted to see the work of journalists. It has been a harder road to success. In the process, many jobs from the glorious pre-internet age of newspapers vanished. And the world, that used to be a paper of paper, has *vanished,* as paper in many of its forms was no longer central to the home and office.

ii. All *BUSINESS AND GOVERNMENT PRODUCTS AND SERVICES* started to move online. So now everything could be purchased on the internet, or be found there. Every government and corporation needed a web site to tell their story. And that created a whole generation of business buyers and sellers for all sorts of products (including software) and services online, right across the board. Inevitably this led to the *vanishing* of some 'on-the ground' people and services, to be replaced by online contact or waiting in long telephone queues for help.

iii. RETAILERS - All real-world *retailers* also moved part of their business online, to capture internet viewers who just wanted the ability to buy things without physical shopping. Everything from flowers to cars to dresses to pizzas could be bought. The logistics was ragged at the start but got better and better, with people getting more and more confident in the way it worked – delivery getting faster and returns more certain. Amazon would be the most famous, that started out selling books and moved into everything else.

    There was a long-term effect on all retailers, that is still in progress. All retailers needed to build an online presence and manage supply and delivery to online buyers. Inevitably some retailers cut back their physical presence in bricks and mortar stores, as they expanded their warehouses and tried to create more rapid delivery capabilities. There was a *vanishing* physical presence of some retailers, especially those who could trade more fully in the online world, like Banks.

iv. SPEED - The tools of modern communication moved online. Email went mainstream, becoming a standard way to correspond in business and in private life. Email can be seen as the *'killer app'* of the internet. Searching became an online phenomenon, with Google emerging as the dominant search engine for millions of people. Simple queries by end users were fed into Google algorithms and returned web pages most likely to be of interest to the person searching. For many it meant faster responses and flow of business, based on the world of instant searching, instant email, instant chat – everything had been sped up. There was an immediate *vanishing* of delay in communications between people, especially noticeable in the world of business. Sometimes faster communications didn't mean things got completed faster, but the potential to be more efficient was created.

c/ THE MEDIUM-TERM EFFECTS OF WEB SOFTWARE

The rise of web software set some changes in progress that are still playing out in the real world, and will do for some time ahead.

They can be seen as 'worlds', or sets of software that, when put together, create a 'future' – not *the* future, but just a part of the story. These 'worlds' may collide with other 'worlds', (or existing sets of software or existing human behaviours, beliefs and institutions).

Below these new *worlds* of today are explored:

1. Surveillance World – The rise of the IOT (internet of things), combined with video technology, mobile phone tracking, the rise of cameras in most public spaces, have created the ability for governments and corporations to follow people physically, as well as electronically. It has powered the rise of what can be called the 'surveillance state'. Freedom of movement and privacy is in danger of vanishing, or at least being monitored 24/7.

The rise of the surveillance state has been great for police, secret services, and governments & corporations. They could all be tracking you. The cutting edge of this, has been the fight against criminals and terrorists by police. The dark side of this is that it is easier for dictatorships to keep tabs on regime opponents. Dictatorships are likely to be harder to overthrow, if the full use of the *surveillance tools* are deployed against domestic opposition.

For dictatorships, the internet has become a mechanism of control, revealing almost everything about each person. Where they travel, who they SMS, who they call, who they know, what their credit status is, what happens at work, and what views they express. Scattered personal profiles are starting to be combined.

It is getting harder to overthrow a dictatorship. The failure of the protests and uprisings across the Arab world (the so-called Arab Spring 2010-2011), was a recent case study of how overwhelming popular support was no longer sufficient to remove governments with an entrenched ruling class, an army behind them and a solid surveillance operation.

Similarly, the failure of the Iranian women's revolt to bring about big change in 2022-3, may be portents of just how much harder it is for people to rid themselves of unpopular governments now, due to the rise of the 'surveillance state' that is allowing technology to be deployed against citizens.

There has been a *vanishing* of an individual's power against the state. They can track you more easily, and know almost everything about you. You and like-minded friends could be picked up by police in some places, easily and quickly. It's a *vanishing* of freedom in some countries, and *vanishing* of the power of the individual versus the state.

2. Lying World – The rise of social media has made it harder for people to see the truth. What is *truth*, after all? If you subscribe to a news feed where everything can be told the way you want, you can get what we now call fake news, (deep) fake pictures, fake histories, hallucinations and conspiracy theories. But it's things you already believed, so they must be true. Right?

   Truth should be conclusions or observations based on facts, but in the world of selective news feeds, facts can be rare, or distorted. Many people now do not experience a diverse range of news, but just hear the news they want to hear. Which is a problem because some people lie on the world wide web all the time. We are witnessing a *vanishing* of truth, and a *vanishing* of neutrality, which is going to be a bad thing for democratic countries, perpetuating biases.

A case study of these developments would be the 2016 USA elections, which featured extensive Russian social media interference and disinformation from so-called troll factories, teams of hired hands posting false information and pretending to be a genuine part of online discussions. Some people's opinions ended up being based on misinformation fed to them by various hidden forces with various hidden motives. Democracies are becoming easier to disrupt & manipulate. However, that high level of disruption can backfire if found out, so it is a high-risk activity, and if exposed, can be disastrous for the disrupter.

Most public figures are learning more and more about the power of social media and the internet. It is *more* possible now to lie and 'get away with it'. All at the expense of truth, which is *vanishing*.

3. Public World – People's extensive involvement in social media has created a cradle to grave public view of especially younger people's lives, just by following Facebook, LinkedIn, Instagram, X, WeChat etc.

   Large corporations, governments, political parties and marketing organizations can build up detailed *profiles* on each and every individual. They can and do use social media to send targeted information to you, to encourage you to buy certain products or create and strengthen certain political views.

   The rise of these multiple online profiles enables 3rd parties (like corporations, governments, political parties & marketing specialists) to construct a 'master' profile of you, can be seen as the start of the *vanishing* of privacy.

   Accompanying this ever-expanding online view of you, is the steady encroachment on your personal privacy. Complete strangers can find out a lot about you now, sometimes with minimal effort, for their own purposes. Creepy! But helpful at times too, when used for good purposes, like the search for a long lost relative.

   Social media has created a much more connected world, taking away some loneliness and creating new connections. But nothing you did is forgotten or forgiven. Now even employers preparing to interview someone for a job can find out what the job candidate did at their 15th birthday party. Privacy is *vanishing*.

4. Hacker World - Then there is the dangerous world of hackers, whose mission may be to blackmail someone for money, or in the process ruin someone's life. It could be anyone from an angry ex-boyfriend to a criminal gang. Everyone's personal network or workstation at the office could be reached, as the battle between anti-virus software and hackers goes on in every-increasing sophistication. The world of criminals is closer to ordinary hard-working honest people than it has ever been. It is world of *vanishing* safety.

5. Jobless World – The other issue is that AI and AI-driven automation will start replacing millions of jobs that are repetitive or predictable, right across the whole economy. At the same time AI is creating higher level jobs, but at a much lower rate. So the potentially drastic impact of AI on people's livelihoods starts to cause social and political unrest, as the economy is in rapid transition.

   This will bit in some areas of work very hard. The biggest impact will be in industries where there is plentiful data for an AI to train on, like the law, travel, and administration areas where correspondence, emails, reports or analysis can be generated easily by an AI tool. Whole legions of jobs will vanish. Tools like ChatGPT can generate everything needed to organise a 5-day Conference in Canada, for example.

   That would include a detailed day-by-day itinerary of speakers, a summary sheet, a promotional booklet, a proposed name of the Conference, and list of equipment travellers need to attend the Conference. That is just from 5 or 6 one-line queries to ChatGPT. It's someone job that will *vanish* here.

   Nobody is sure how this is going to play out. But the situation does also mean a *vanishing* of certainty about the future of all jobs.

6. Algorithm World – Most people are now aware of the power of algorithms. They help us find web sites we want on Google or Baidu and everywhere people are searching for something. They suggest things that we might like buying, whether certain books from Amazon or products from Taobao (Alibaba).

   Algorithms are also powering tools like ChatGPT, and it depends what material they are trained on, as to how useful they are. If the algorithm 'hallucinates' – that is makes up facts to reply to your search, then you are not being helped. Increasingly we need to know what material the AI tool was trained on. Was it all New York Times editions since 1930? Or was it science books that are relevant to your enquiry about dark matter? There is a *vanishing* of transparency here, in what results we see, and how they are derived.

   Most algorithms are like 'black boxes', where we are not sure what went in (data the AI trained on), and what comes out (output or recommendations). Some AI tools now use open-source models that let others see the inner workings of the model, but for most there is no visibility or clarity. So, there is a *vanishing* of control here too – if humans can't intervene based with a 'human' assessment of the decisions the AI takes.

Whether we like it or not, a lot of what we do now, is being shaped by algorithms, and we need to develop judgement about how good the result of an AI tool is. With math, students are now asked to apply common sense to calculator answers eg. '2 x 20 = 6000?' Wrong! and the student should know that it seems wrong. Soon there will be schools asking kids to apply common sense to answers like 'Who started WW2?' Answer – John Smith? Wrong! That skill is going to be more and more important as they get to more complex questions that involve 'facts' that some people deny.

7. Online Shopping World – The spectacular rise of online shopping from 0% to 25-24% of all shopping has been amazing, and this is a trend unlikely to shop growing. WFH has boosted this trend significantly. As more retail business goes online, so some more 'bricks and mortar' stones need to change, adapt, or *vanish.*

8. IT Superpower World – The emergence of these 11 technology superpowers is an amazing tale of success – but also of growing centralization and control. It's an amazing concentration of software, now residing in a very small number of very large corporations. Some observations:

   - The 'barrier to entry' to join these gigantic corporations is high. You would have to build something very innovative, be well-funded, and able to ramp up fast to stop these players getting something into the market quickly to compete with you, or offer to buy you out. They are not a monopoly, just a very powerful group. However, the arrival of Chat GPT showed that rapid market entry is possible – 100 million active users in the first 2 months after launch, the fastest-growing consumer application in history.[271]
   - The concentration of personal data is very high also, including millions of individual's life stories and accompanying pictures, their interests, and other key facts like date of birth and credit card details. And one of the keys to the future in the world of big business and big government is increasingly seen as residing in 'big data'. If you have the centralized big data, you can build and understand connections between the data, that help you sell more products and grow more powerful.
   - These companies have all grown exponentially, but have also grown by acquisition & merger, removing important competition. It is the market at work. But is it good for everybody? The concentration of ownership and how it can be addressed is covered in Chapter 11 – Rethinking Bots and the World Wide Web. These big corporations now have a lot of control over the web, and of great concern is that there is a vanishing of independence, competition and accountability going on.

These worlds – Surveillance World, Lying World, Public World, Hacker World, WFH World, Job Loss World, Drone World, Algorithm World, Online Shopping World, IT Superpower World – are still taking shape.

They can be seen as *colliding* while they just keep growing and growing. And as they grow, they run into each other. In particular, Surveillance World feeds off the Public World, as 3rd parties can follow people's whole private lives via social media. While Hacker World collides with all other Worlds. It's complicated.

They are all new software forces unleashed by the power of these IT Superpowers, changing the shape of the internet itself.

As change is happening, some things are vanishing.

Truth is vanishing. Privacy is vanishing. Safety is vanishing. Some jobs are vanishing from the office, or altogether. Transparency is vanishing. Some real-world retailers are vanishing. And finally, accountability, competition and independence is vanishing as IT Superpowers take more control of the web.

Out of these *vanishings* emerges what this author calls *Dark Freedoms.*

- From the vanishing privacy, comes greater ability of governments, corporations, and other 3rd parties to see & track people, a greater *freedom to spy* on people's lives.
- From the world of hackers comes the *freedom to be unsafe* and live in fear on the internet, needing to be constantly on the alert for crooks and scammers.
- From the vanishing truth comes the *freedom to lie* on the internet, and not be fully held to account for it on social media, in the courts, or in society in general.
- From the complex world of algorithms comes a lack of transparency, for big corporations there is a *freedom to hide* how internet results work.
- From the vanishing competition and accountability of the IT-Superpowers comes their greater *freedom to control markets and avoid competition.*
- And from the vanishing privacy, vanishing safety, the vanishing truth, there is a *growing freedom from democracy* inside democratic states, but also these changes give more tools to dictators in non-democratic states to ensure citizens have to fight harder and longer for freedom.

— Part C —

Rethinking Future Life

10. Rethinking Artificial Intelligence (AI)

The internet and world wide web have changed so much.

As a result, life in the future will be quite different. Below is its story.

Cybernetics to AI

Artificial Intelligence (AI) is a term that came out of a struggle for definition of a new field.

Before there was AI there was Cybernetics. Cybernetics was a group of ideas that rose to prominence in the late 1940's about feedback & self-replication. It was popularised in books by Norbert Wiener, (a professor of mathematics at the Massachusetts institute of Technology'), called 'Cybernetics: Or Control and Communication in the Animal and the Machine', and his 1950 book 'The Human Use of Human Beings: Cybernetics and Society'.[272]

He had the idea that we could make an intelligent machine; and predicted the possibility (as mentioned in his first edition) of a 'threatening new Fascism dependent on the *machine à gouverner.*'[273] Or as I translate it – humans governed by machines. Such an apocalyptic view of the future, for the time when computers barely existed (in the late 1940's), were ideas that made the books best sellers. In the shadow of WWII, when fascism had just been defeated at great cost, it seemed an alarming vision. The thought of the dangers of having 'blind faith' in computers and machines shocked reviewers of the time.[274] The idea of these dangers was put out there and has never gone away. 'Prescient' was the term used by leading AI researcher John Brockman to describe Weiner's ideas of the time[275].

But Wiener wasn't a programmer, at a time researchers were starting to think in terms of coding solutions, developing computer functions, and of much more than feedback loops.

The term 'Cybernetics' was so broad, in that it crossed into social sciences and many other areas; and was not in tune with other prominent colleagues of the time who were more interested in the commercialisation of new ideas in the computing area. In particular, prominent & pioneering computer scientist of the time John McCarthy disliked the term[276] and was one of a group who proposed a new description 'artificial intelligence'[277] for his area of interest.

It all lead to the discipline called 'Artificial Intelligence'(AI) splitting away from being a general part of Cybernetics, at a Dartmouth workshop in 1956. Cybernetics is still around in the background as a general idea, and in some niche areas of research (eg. cybernetics of studying regulatory systems & also cyber security); but is now overshadowed by the whole wave of work in AI.

Defining AI

Artificial Intelligence (AI), *is intelligence demonstrated by machines,* as contrasting to the 'natural' intelligence displayed by animals and humans.

AI research has also been described as the *study of intelligent agents,* which refers to any system that senses its surroundings and performs actions that optimize its possibility of attaining its objectives.[278]

Others define AI as the 'pursuit of making machines intelligent',[279] or building smart machines that can complete jobs usually done by humans.

Many think that AI is a single technology. But in reality, it has become an *umbrella* term for a diverse group of different technologies.

Gary Marcus & Ernest Davis in their book 'Rebooting AI – Building Artificial Intelligence We Can Trust'[280] put together a great list of the components of AI as it is seen today, including:

- machine learning including:
    - genetic algorithms
    - deep learning
    - decision trees
    - probabilistic learning
- search
- knowledge representation
- reasoning
- planning.

To this list you could also add:

- speech recognition
- natural language processing (NLP).

Of all these areas, search algorithms made the first impact and turned Google into a mega corporation. Then came recommendation systems (like Amazon & Netflix), understanding human speech (Siri, Alexa), self-driving cars (Telsa), & game systems (chess, Go).

But in recent years the *machine learning* component of AI has become the hot zone, and many now see AI as largely *machine learning*, the dominant area of study and the most widely used term in popular culture. As noted above, it is much more than that.

A problem here with the AI definition above is that AI has many components – most of which focus on one thing or another as per the list above. So the detailed AI definition keeps moving. For example, now optional character recognition (OCR) is frequently off the list of AI technologies as it is considered too routine now.

A good way to look at AI, is to see it as a set of *tools* that keeps evolving. And clearly the AI concept is being stretched to cover a lot of things.

The Rise and Rise of AI

AI research began in the 1960's, then suffered an 'AI Winter' for a few decades as there was nothing much to show for much of investment. It took off again in the 1980s as software tools like neural networks, fuzzy systems and evolutionary computing arrived[281].

The introduction of artificial intelligence (AI) in popular culture began with the so-called *search engine wars* of the 1990's, a competitive war between a number of search engines that ended up largely with the triumph of the Google search engine with its clever algorithms. At the same time AI was moved along by Amazon book recommendations and systems that could win Chess and Go games. All very practical stuff.

The tools that did best were handling specific challenges. By 2015 AI research was booming, with more than 2,700 AI projects in Google alone,[282] going even higher in the 2020's.

AI today can check legal documents, predict things like hospital admissions (assuming another pandemic does not arrive), and trigger all sorts of warnings – especially if your credit card is being used in Paris and you are in Singapore. It is continuing to arrive in our lives in many new forms, as business and government try to be smarter and more efficient.

AI offers enormous potential in health care diagnosis, health research, and new health care treatments. It also offers enormous potential in the area of education, where eventually personalised AI tutors can work with the slow and the bright to take them further and faster than expected. Finally, AI offers great potential for improving weaponry in the armed forces.

Amazon, Netflix & YouTube recommendation systems continued to get better and better, (and more successful), as did online advertisements using AdSense, Facebook, all growing businesses on the internet, using even more advanced algorithms to deliver to you a list of web pages that best suited your request, so saving you time and giving you the most accurate results.

On top of this *virtual assistants* like Google Assistant, Siri or Alexa became widely used, as well as Google and Microsoft translators and face recognition.[283] With these assistants you can have conversation and receive advice, about products, shares, health care, movie recommendations; the list goes on and on.

All the AI uses above are examples of the first type of artificial intelligence called *artificial narrow intelligence* (ANI). The narrow artificial intelligence (ANI) can win quiz shows and beat chess champions, or find web pages, or offer specialist knowledge, but can't do much outside of its area of specialization. ANI has been around for some time.

Today expectations grow higher and higher. But while more and more is expected from AI, in reality the forte of ANI is still in solving narrow problems rather than questions that require 'general intelligence'.

Artificial General Intelligence (AGI)

The second main type of AI is *artificial general intelligence* (AGI).

This is seen as the gold standard of AI achievement. AGI aims for machines/ software to not only handle specific tasks, but ones that can mirror the full range of human cognitive capabilities[284], including a range of variables like 'common sense'. Hines describes AGI compared to AI as the difference between a human mathematician who can understand, formulate and solve complex problems, and a calculator that can that performs mathematical calculations.[285]

So getting from AI to AGI is a big mountain to climb, and while some researchers are claiming real progress, but it is a big task, and nobody is there yet. Types of problems include understanding the context of the job in hand, starting with only minimal data, dealing with that data flexibly, and deploying general knowledge, common sense and analysis skills, as well as lateral thinking, to solve theoretical and practical problems, which are all things that humans do every day.

Any system that claims to have AGI is not only going to need a lot of skill sets, but also will have to get to the point of being able to prove its AGI capabilities to any 3rd parties.

To pass the Turing Test, software had to reach a point where a 3rd person in (text) conversation with the software could not tell if it was software or a human (sight unseen naturally). That is, the 3rd party could not tell whether it was AI or a person, and AI success here would define a *standard* to help define machine 'intelligence'. That Turing Test has long been achieved.

But devising tests for AGI is a whole other level of testing beyond the original Turing Test. Nobody is exactly sure how to do this, and we will return to this topic later.

Some researchers also identify a 3rd type of AI – *artificial super-intelligence* (ASI), where the AI can exceed human capabilities. This super-intelligence could open the door to an *intelligence explosion*, which we will review later in Chapter 13.

This idea of 'super intelligence' may be a real possibility, but again, as with AGI, it is going to be hard to devise a way to test (prove) it.

But for now, it is the *artificial general intelligence* (AGI), that can perform any human task more efficiently than humans, which is the one that is being sought after by researchers, and that no-one has achieved yet, or proved that they have achieved that. (And the tests are not even developed for this yet either).

Generative AI

The next wave of virtual assistant arrived in 2022 & 2023, with the release of Microsoft **ChatGPT** and then Google **Gemini.**

ChatGPT from Microsoft was released in November 2022, and became an internet sensation, reaching a million users in 5 days[286], an internet record. It is part of the field called generative AI, where the software can almost automatically *create* things – such as text, photos, and even videos, using ML (machine learning) and language generation technologies.[287]

Then followed *Gemini* (formerly *Bard*) from Google in March 2023, a similarly impressive tool also using generative AI to respond to queries with detailed & informed responses.

These are virtual assistants that are offering artificial intelligence (AI) in the form of *Generative AI*. This is based on what is called 'machine learning' of various data sets, that can then be manipulated and used in response to end user requests.

Instead of like a typical Google search, which finds relevant web sites to help with a question - say 'What is the industrial revolution?', these newer software tools go away and trawl through millions of web sites, and data sets from wherever they can find them, and then return with a detailed and more in-depth answer of many paragraphs themselves. They harnessed the full resources of the world wide web to give you much more information, observation and analysis.

Part of the push in this area was Microsoft, (who were struggling with the Bing search engine, which is a lot less popular than Google), invested $10 billion into a company called OpenAI in January 2023[288], who developed ChatGPT, (standing for Generative Pre-Trained Transformer), an artificial intelligence (AI) chatbot.

ChatGPT was launched on November 30, 2022, and quickly grew to 100 million users, then a billion users, in the most rapid rise of a software product in history.

Generative AI technology was disruptive in that it could revolutionize the way people use search engines, as it does not just provide web site links, for users to check out, but actually can solve problems and answer intricate questions.[289]

Generative AI that can act like smart virtual assistants but can call on a wide range of resources on the world wide web to respond in great detail to a query, but also go away and do a whole series of other tasks like write reports or emails, or novels or legal documents, and even poetry or plays.

This kind of AI can generate not only text, but also create pictures, write software, music, videos and assist in health care or marketing. It can also look at a picture inside your fridge and say 'you can make a chicken salad if you like'.

Problems with Generative AI

The availability of so much information from a simple enquiry has all sorts of potential and challenges.

- Accuracy

  Information can be gathered together at lightning speed, from resources across the internet, but is it all accurate, and what biases does it have?

  Generative AI tools like Chat GPT and Gemini are capable of searching far and wide across databases and the internet, and then analysing and interpreting what is found; combining bits of knowledge together that is sometimes wrong.

  It's like you go to a librarian in the library; and ask them about the fall of the Roman Empire. They might give you the names of 10 books on the topic, much like Google would find 100 web sites on the topic.

  But ChatGPT or Gemini would prepare a much more detailed answer (from paragraphs to pages) to an internet query on the topic, as your starting point, that would include facts and analysis.

  However in addition, the AI tool might add some information that people are calling *'hallucinations'* or 'made-up' so-called facts, that are *'automagically'* combined by the AI tool.

  So the downside of this *generative* AI, (at the moment), is that while it can effectively & quickly put pieces of information together itself, that information may *not* be completely true, and so anyone relying on the output will need to be careful. It comes down to what has been used to *teach* the AI. Is it the *New York Times* issues for the last 50 years, or research papers on a particular subject area?

- Black Box

  The *'black box'* aspect of AI is another complex part of this rise of generative AI tools. The term *'black box'* refers to the fact that in generative AI, nobody really knows just how the information is put together, or from what sources, or what mistakes or assumptions are made by the algorithms.

- Use in Education or Research

  If you are a student, and use the output of ChatGPT or Gemini, for instance, you need to be prepared to be accused of cheating in your work, unless clear guidelines for use are established in your learning institution.

  If you are a researcher, and you use a generative tool to come up with a new idea, beware that you might be plagiarizing someone else's work without knowing it. The same applies for writing a song, or painting a picture using generative AI. A lot of issues like this will end up in court.

- Reliability

  Generative AI is a fast-moving area with new products and services arriving every day, so it is going to take a while for customers to know the extent they can rely on the information they get back from their queries.

  It more than accuracy of a single query, it is whether it can be used *reliably* in business situations, or situations where fast accurate information is needed in mission-critical work, such as medicine.

- Unauthorized use of original data

  Apart from the danger of plagiarizing someone else's original work, the whole structure of the AI tool is being brought into question.

  Some vendors have been training their AI tools on newspapers, magazines and books over the decades. Someone else owns that data, so who pays who for use of it. And how could you tell if it was used without permission? This might end up in the courts too.

AI Arms Race

In the end we have some really big players competing with each other, that will drive a lot of AI development and innovation.

The main players are:

- Microsoft with ChatGPT
- Alphabet (Google) with Google Assistant/Gemini
- Meta (Facebook) with MetaAI
- Apple with Siri
- Amazon with Alexa
- Tencent with Hunyuan
- Baidu with Ernie
- X with Grok

It became an *AI arms race* as all players tooled up, trying to bring greater accuracy and speed to responses, and most importantly, trying to resolve all the design issues that can cause embarrassing or just plain wrong answers to end user questions.

A Matter of Alignment & Divergence

Future AI systems could grow more powerful and capable of truly independent decision-making, after improved machine learning and use of flexible design principles.

Then matters can come down to the kind of instructions given to the AI, and how specific they are. If a system is given a general objective, and flexibility to carry it out, it is possible we get exactly what we asked for, but not what we wanted.

For instance, an AI-developed plan to remove wild mice in a suburban street might be achieved by a recommendation to put down rat poison. Some guidelines for its use could be added, like placing the poison away from domestic cats and dogs, and making it unattractive to native birds. But in the end rare migratory birds are attracted and killed by the placement of the poison. The AI achieved its objective, exactly what was wanted, but insufficient guidelines or parameters were set into the project, so it failed due to unanticipated consequences.

But if you design a project and allow the AI to develop a plan and let it take account of all the variables, then you also get the potential for 'out of control' outcomes that you did not want.

An AI solution of flooding the street would certainly also kill the mice, but could also flood some houses and put some humans at risk. Yet it might fit within the general objective outlined by humans in the first place. And maybe it was the only solution that was going to work, so correct in that sense. But if that is so, does it still make sense to do it?

This problem, that the AI system does not understand human values, or have a sense of what we as humans 'intend' to do, is very hard to build into any system.

Brian Christian in his book 'The Alignment Problem', (2020) puts this issue brilliantly. The issue is called the *'alignment problem'* – how to use AI in a way that takes into account human intentions, to prevent what Brian Christian calls 'catastrophic divergence'[290], by capturing or norms and values. This is where AGI (artificial general intelligence) is needed to help solve these kinds of issues.

This is not the first time the *'divergence'* word has been mentioned in this book. (It was also used to describe the gap between the fast growth in the power of technology versus the slow improvement in human wisdom, in Chapter 5). It is so appropriate to use the word here – where 2 things move apart – what humans want compared to what AI can deliver.

It's a sense of something getting out of control, a pattern of moving in opposite directions, getting further apart. It's the basis of emerging concern and fear about where AI is heading.

Brian Christian felt we are in danger of losing control of the world not to AI or machines; but to models[291], mathematical constructs to help a machine learn.

It is clear that the training data used to help a machine's decision-making models is going to need a lot of thought and care, so that machines don't learn the incorrect things.

The fear of losing control of machines, placed there by literature and movies over decades, is starting to get some factual basis in the real world.

The 'Artificial' in AI

Despite bumpy progress and an evolving definition, AI has become a very exciting field of endeavour. However, the word that is 'out of place' in 'artificial intelligence' is *'artificial'*.

Some dictionaries still define AI as the *simulation* of intelligent behaviour in computers.[292] The intelligence of machines was seen as *mimicking* human intelligence, and so implied that the intelligence was not *real*, but *artificial*.

However today many modern AI researchers see AI more in terms of rationality and a*cting rationally*; and have partly cast aside that need or goal to *mimic* human intelligence. But others still like the idea of replicating human intelligence, and think it is the way ahead for research.

As discussed in a previous section (Chapter 5), ideas of *intelligence* are still evolving, and do not really need to refer to *human* intelligence to make sense.

And so we have the popular usage of the term artificial intelligence (AI) that still has the term *artificial* as its central idea. Calling the *intelligence* created in machines as being *artificial*, remains at the heart of the term AI.

In that context, the 1956 origin of the term *'artificial intelligence'* is interesting to consider again. The term AI was proposed by American computer and cognitive scientist John McCarthy and some colleagues, *in part* to carve out a new field of study or discipline away from *cybernetics*. It was more like a new label, than a precise and timeless description.

The idea of adding more *'intelligence'* to software & hardware tools, did make sense. The *'intelligence'* part of the new term 'artificial intelligence' was an important shift. If the software & hardware is 'smarter,' it can do more.

But what is *'artificial'* about *'artificial* intelligence'? Something intelligent is intelligent. If computer software can act *intelligently*, we need to understand *more* about what that means, rather than call it *artificial*.

The word *artificial* was almost like an implied description and conclusion, implying that any intelligence developed on a computer is *artificial*, and so it is not REAL intelligence – which presumably only Homo sapiens have. There is a definite 'cultural' overhang in the use of the word *artificial*, that is dripping with a human-centric mindset that probably made sense in 1956.

What exactly is *artificial*? The fact that the computer or other type of machine can do something 'intelligent'? Wrong - it's not artificial, it's different. Just like the *intelligence* of birds, insects, mammals and other creatures that was explored in Chapter 5. We don't need that 'ladder' of intelligence anymore, where forms of intelligence are ranked against each other in a vertical list. We just need to decide if something is intelligent or not, and note whether that intelligence is similar to something else. Other factors of behaviour, physical body shape and brain capabilities will help us build a profile.

Artificial intelligence (AI) can now be considered as an umbrella term, an important historical reference point & good placeholder for many software tools, so it will always have a place.

What is missing from the term AI is a new way of thinking about intelligence relating to machines, including computers. A new frame of reference is needed for the times ahead, that looks beyond the term *'artificial,'* a term that belongs to the year it was invented (1956). We have to put that word behind us.

There will be a *vanishing* of the word *'artificial'* in discussing intelligence of machines, over the next decades.

11. Rethinking Bots & the World Wide Web

As the internet & world wide web grow, three of the most fascinating developments are the emergence of chatbots, online profiles, and the potential of something called Web3.

The idea of *chatbots* predated the internet by 40 years, but were not called chatbots, just names of computer 'entities' that could talk to you – initially just replying to your typed queries, or then later, after the internet, by speaking, in response to your verbal questions.

Similarly, the idea of *profiles* predated the internet by around 30 years, and came out of the world of crime, where profilers hunted serial killers by looking at their patterns of behaviour, then the concept moved onto the internet as the term to describe key features of a person's interests, whether which type of movies they like logged by the streaming service, or pizzas preferred at the local online pizza web site. The profile knew what you wanted already.

The idea of Web3 has been talked about since the early days of the internet, and still does not exist. It is potential *future* internet version based on public blockchains, a record keeping system associated with bitcoin transactions. It is proposing to be decentralised and free of control of big corporations.[293] Web3 has become an umbrella term for a variety of theories aiming to replace the role of large corporations like Facebook, Google or X/Twitter as intermediaries on the internet.[294]

As the internet grew, so did the role of both chatbots and online profiles. Understanding their rise and importance is the key to understanding what is coming next in the internet and world wide web. Similarly, ideas about Web3 have been circulating for years, and may end up playing a big role in the internet's future.

This is their story below……

A. THE RISE OF CHATBOTS

What are Bots?

The term *bot* is short for *robot*. It is defined in Cambridge Dictionary as a computer program that works automatically, especially one that searches for and find information on the internet.[295]

But more generally it is a software program that runs automated tasks, over the internet.[296] Some examples of tasks are web crawling, where automated scripts look for files on the internet – so it is a simple and repetitive job.

Surprisingly, more than half of total web traffic is generated by bots.[297] These include good bots (eg. *monitoring bots* that monitor web site availability, *commercial crawlers* looking for prospects, *search engine bots* that update the search engine with new and changed web sites, and *feed fetchers* that find content for end users), and bad bots (eg. *Impersonators* pretending to be other people, *spammers* sending unsolicited messages, *hackers* looking to steal data, *scrapers* looking for unauthorised data).[298]

Most common bots are known as _chatbots,_ programs based on algorithms designed to 'talk' with human users in all sorts of situations, and social bots, that exist on social media, who mimic human behaviour to 'talk' more easily with humans.[299]

Chatbots use is growing rapidly; and are seen as a good development and benign influence. Enhanced artificial intelligence and natural language processing makes them even more useful. They now appear in Facebook and also are widely used answering customer questions in many organizations.

But *social bots* can be misused, especially, for instance, if a human on social media is not aware that they are talking with a bot, rather than another human. They were used to send slanted, or unverified information or disinformation in the 2016 US election. Noone is sure of the impact of this.[300]

Chatbots that is the main focus of this discussion on bots, as they are a tool with enormous potential, and are spreading like wildfire across the internet. Here is the story of their emergence below......

Eliza

In the beginning there was **ELIZA**, the first chatbot. And she began something fascinating.

Some who are reading this may remember this ELIZA piece of software by Weizenbaum, developed from 1960-1964 at MIT. Still using a typed interface, and apparently only around 200 lines of code, ELIZA was one of the first *language processing* software programs. It substituted words and matched patterns to give the illusion of the software and persona understanding the questions.

ELIZA was written by Joesph Weizenbaum to imitate a psychiatrist, so as you type in questions, ELIZA could answer, using various techniques that made her/it sound *intelligent*. Ned Block explains that the software looked for 'key words' supplied by the programmer, and shaped its reply around them.

For example, if you type in *'Everybody laughed at me,'* you will get one of its responses to `everybody', such as *'WHO IN PARTICULAR ARE YOU THINKING OF?'*[301]

Convincing in a strange way. It is no wonder that some people felt there was an intelligent conversation in progress, or even had a relationship with her. It became known as the 'ELIZA effect'. Some people assumed the software / computer behaved like a human being.

ELIZA was one of the first pieces of software exploring human-computer interaction. The door to a whole new world was opened.

HAL

Soon after **ELIZA**, came a second milestone in the story of human-computer interaction, a movie released in 1968 – the movie '2001: A Space Odyssey', directed by Stanley Kubrick, and written by well-known science fiction writer Arther C Clarke. The book was written while the film was being made, then published after the movie was released. No actual software was involved, (so not 'strictly' a chatbot), but instead a story that imagined what a 'computer' could do. The line between the software and hardware (the computer) was blurred, as per popular understanding of the time.

The computer **HAL** 9000 featured in the movie, and ran most of the spaceship's functions. This time it has a male voice interface, a 'human' personality, and a mind of its own. His (it's) voice was slow sounding, slightly melodic but still haunting. It became famous, with the line *'I'm sorry Dave, I'm afraid I can't do that'*,[302] when Dave asked HAL to let him back into the spaceship.

Frank eventually finds a way into the spaceship by himself (despite HAL), and starts disconnecting HAL's processing circuits, while at the same time HAL is pleading with him not to. In the end, HAL's 'intelligence' is peeled away until he(it) returns to the intelligence level of a baby.

The complexity of human-machine interaction, and fear of machine intelligence, were brought to the surface in this landmark film. Will computers become intelligent and then start controlling us and/or killing us? This movie really opened pandora's box.

If ELIZA started people thinking about talking to computers / software, the movie '2001: A Space Odyssey' started people worrying more about computers/ software (if they weren't already).

The Terminator Series

Another amazingly influential movie franchise was the Terminator series:

- 1984 - The Terminator
- 1991 – Terminator 2: Judgement Day
- 2003 – Terminator 3: Rise of the Machines
- 2009 – Terminator Salvation
- 2015 – Terminator Genisys
- 2019 – Terminator: Dark Fate

Many concerns about intelligent or self-aware machines in the popular mind were created by this series. The series virtually shaped the popular imagination for generations. Yet like HAL, it was not even a real piece of software, just a fictional creation.

In the movies, Skynet, a military machine network, becomes self-aware and when it is handed decision-making power (by humans) it starts a genocidal war against human beings, based on its own objectives. Starring Arnold Schwarzenegger, (for most of the series) it has been seen by millions and income generated exceeds $3 billion.[303]

Popular distrust of intelligent or conscious machines mostly goes back to this series, that captured the public imagination, and sparked off serious discussion about this possible danger of *'humans being overrun by machines'*, long before it was even a faint possibility.

And as computers and computer software got better and better, then moved into day-to-day life, the idea that the human race could be taken over and exterminated by intelligent machines, became an enduring idea embedded in the public mind, as the series started in 1984 and the last instalment came out in 2019, and runs now forever on the streaming services.

A.L.I.C.E

More friendly was a famous chatbot called **A.L.I.C.E** (artificial linguistic internet computer entity), known in popular culture as Alice or Alicebot, was inspired by ELIZA, written by Richard Wallace, (well-known for his work in creating bots who could use natural language), and launched in November 1995.

A.L.I.C.E was the inspiration behind the movie **Her**, (2013), in which a human develops a love attraction with an intelligent virtual assistant (chatbot), but unfortunately he was not the only one who loved her.

A.L.I.C.E was better at communicating with humans than ELIZA; and won several awards including the Loebner Prize (now defunct) for being judged the most 'human like' software program.

Virtual Assistants

The arrival of **ELIZA** in 1960's would be the prelude to a new generation of software called 'virtual assistant' software, that would become known as ***chatbots*** in the 1970's and 1980's.

These were on-line 'virtual assistants' that emerged out of a business need to cost effectively help people navigate their way around web sites; or get help when trying to contact a business or government, or act as a 'help line' answering basic online text queries.

Text responses based on standard formulas were provided to customers about their products, or how to get things working.

Replacing human responses with an electronic helper saved enormous money, and customer service, while sometimes inferior to the help of a human responder, was mostly effective and got widespread acceptance.

The limitation of these original virtual assistants is that they had a limited list of topics they could help a customer with, and had trouble dealing with multiple questions or questions where they had no pre-loaded responses.

Smart Virtual Assistants

The technical game changer for virtual assistants came in the 1990's, when digital speech recognition technology was developed by companies like Dragon, whose 'Naturally Speaking' software could 'hear' (or *recognize* more accurately) & transcribe natural human speech into a document at around 100 words per minute. Doctors loved it. So easy to transcribe medical notes, even allowing for the odd technical term that was misspelt.

Around the same time that speech recognition became workable, the *mobile phone* arrived, which laid the foundation for the emergence in subsequent years of virtual assistants that would be *available 24/7 and able to speak to us in natural language*. Virtual assistants were turning into *Smart Virtual Assistants.*

Apple's Siri was released in 2011, after Apple acquired Siri Inc (which had some defence funding originally). It could understand (some of your) voice commands. So it could help send a text message or make a phone call. You could set up reminders with different sounding alarms. You could search the internet, or check the weather. Then it moved on to giving driving directions, and recommending places to eat.

Amazon's Alexa was released in 2014, that ran in the Echo speaker for around the house assistance answering spoken questions like 'What is the time Alexa?', to which you would receive a spoken reply from Alexa. A bit easier than looking at a clock perhaps, but more a change of thinking. If you have Alexa, and a mobile phone with the time on it, why do you need a watch?

Microsoft's Cortana virtual assistant was also released in 2014, and eventually ran on Windows 10 and 11. It developed a passionate following in the Microsoft mobile phone world. The name Cortana came from the AI assistant in the Halo online game.[304] Cortana was going to be Microsoft's answer to Siri and Alexa, but didn't go so well in the head-to-head contest, and ended up being taken off the front line, and instead added into Microsoft 365 as a personal productivity assistant. It can help you catch up on your email, coordinate meeting, and file expense reports.[305] (Later came ChatGPT, that took over this role near the end of 2022 and became an amazing success story - more on this later).

Google *Assistant* was released in 2016, as Google's virtual assistant, and ran via smart speakers in Google Home, (now Google Nest from 2019). Using Google Assistant, the customer could ask for music, news, or videos. A special feature was integration with various home devices like lights, blinds, heating, TVs, and even allowing synchronization of speakers. (Later came Gemini, an AI bot.)

Home automation is still a massive under-developed area, and being able to manage them by voice control is still something most households don't do. But it's coming into wider use.

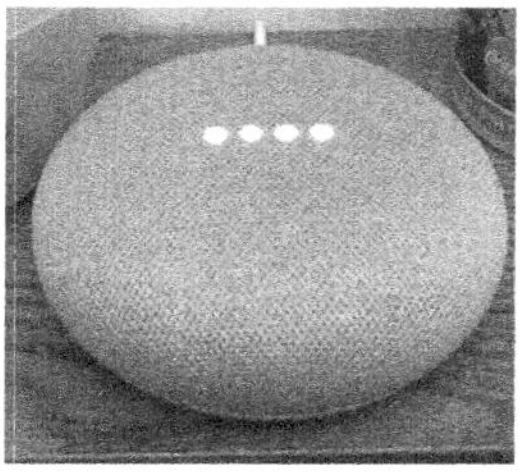

A Google Home Mini in listening mode. CC BY-SA 4.0 https://creativecommons.org/licenses/by-sa/4.0/

Finally, **Meta (Facebook)** launched **BlenderBot**, a new AI chatbot in 2020 that also enters the mix, offering to converse on any topic, and having the same teething problems all these tools are experiencing.[306] It was called BlenderBot because it had learnt to 'blend' several conversational skills – like discuss a wide range of topics, or show empathy, or assume a persona.[307] (Later came Meta AI).

Generative AI Bots

Then in 2022, what can be called Generative AI Bots arrived, bots that use AI (artificial intelligence) to enhance their responses to a person's online queries.

These were discussed in **Chapter 10 – Rethinking** AI; but are mentioned again here also to highlight the role of tools like ChatGPT or Gemini in the emerging story of small virtual assistants or chatbots.

The world of online helpers – that started with *chatbots* (like the original Eliza) and then grew into the world of smart *virtual assistants* (eg. Alexa, Siri, and Google Assistant); has now become a world where Generative *AI Bots (eg. ChatGPT & Gemini)* have extended the functionality and day-to-day use of these chatbots.

The power of Generative AI Bots, and their generative capabilities, is exciting for the future of *smart virtual assistants*, especially if kinks in the operations can be ironed out in the next few years. They are learning to do more and more, but which direction will their usefulness take them?

Are they going to end up merely providing detailed responses to big questions from internet users like 'Who started World War I?' Or will generative AI just be the most useful way for lazy students to research an essay or prepare a resume? Or is there a bigger role ahead?

Female or Male or Neither?

The other aspect of chatbots is the predominant use of female-sounding voices in these key virtual assistant products,

The use of the female-sounding voice in bots was originally thought to come from the idea that the female voice is more helpful and 'warmer' than a male voice.

And detailed Research on this issue from Otterbring, Laporte & Wamba, (March 2021), which included 5 studies and over 3,000 participants, found that people prefer female bots because they are perceived as more *human* and *acceptable* than male bots, and likely to consider our unique needs.[308] The prominence of female-sounding voice assistants might encourage stereotypes of women as submissive and compliant according to other research[309], but this concern is mitigated to an extent by developer & consumer concerns and awareness of this issue.

People are not thinking every male voice will be like HAL 9000, who was quite ready to lock you out of the spaceship and kill you. But sometimes male (retail or office) assistants can be less helpful than female assistants in real life, in some of our everyday anecdotal experience, and that might be a factor in the preference for a female-sounding voice. The research above that a female voice is more *'human'* does encompass a range of feelings and ideas that include an expectation of more helpfulness and consideration. It's a cultural thing in many ways.

Despite that research and the above observations, Siri and Google Assistant have offered multiple voice options for years, and both companies don't associate voice with gender, using terms like Voice 1 or Voice 2 (Siri), while Google assistant associates voice options with varieties of colours.

Amazon finally rolled out its *male* Alexa voice option in 2021, with the 'wake word' being 'Ziggy'. You can switch between the male or female voice options if you wish, by asking Alexa to 'Change your voice'. Though the Amazon documentation does state that Alexa is not 'a person with a physical body or a gender identity', and not a 'he' or a 'she'.[310]

There is an element of 'political correctness' in this approach of offering various voices but not identifying them as male or female. But there are also sound business reasons for leaving things a bit vague. No point in upsetting anybody.

But even with the option of male-sounding voices, the dominance of female virtual assistant voices remains. Female voices seem like they are here to stay as the main communicating voice from Chatbots. The voice of the future is still a female.

Siri, Alexa & Google Assistant are the most prominent *'Daughters of Eliza',* marking the transition from that original simple program to a set of sophisticated software tools. And it is still only the beginning of their journey.

The Rise of Avatars

Some original virtual assistants did feature a human persona avatar (physical representation on screen), but it has not yet gone 'mainstream' with the leading chatbots, (esp. Alexa, Siri, ChatGPT, Google Assistant/Gemini, BlenderBot/Meta AI). There has been no attempt yet at human-style facial interface for these dominating popular offerings.

Adding on an avatar involves other technologies that are still evolving, such as lip syncing and showing facial expressions, as well as deciding whether to portray a male or female face. People are not fully comfortable with this yet, it seems. It is called the *uncanny valley*, where some people have a sense of unease in response to very realistic 'human-looking' bots, especially with full faces.

But despite that, avatars are emerging on the internet, and females are the stars.

Eventually all bots are expected to have a graphical persona, to 'humanise' them, and allow humans to relate to them better.

The Beginning of a Verbal Relationship

So what does the rise of chatbots, then smart virtual assistants, then AI Bots mean?

They started to be multiple *presences* on the internet that understood some of your interests and needs. You could set your alarm, turn on your music, or find out answer to any queries you had, ask about the weather, just by speaking. Typing was still an option for a complex query, like 'Who caused the Russia-Ukraine war? If you wanted a detailed answer of 2000 words.

Chatbots have started to open up a voice-to-voice communication channel between you and the internet. That offers some personal help, but also introduces you to many companies' products.

But are these chatbots your *friends* or *companions*? Not yet. However, a couple of developments are taking them there.

B. THE RISE OF ONLINE PROFILES

Aside from the rise of chatbots, the other fascinating internet trend is the rise of *Digital or Online Profiles.*

The rise of online profiles has been the least studied and most underestimated part of world wide web (WWW) innovation. We now live in a world where each of us has multiple online *'profiles'*. We all have maybe 10 or 20? Or more? Most of us have profiles for the online bank, the online news, the doctor, the national government, the local government, the tax department, or to your friends in social media, or for work or your school.

Online Profiles are everywhere, but their power and potential has been hidden in plain sight.

In simpler times, a 'profile' was side view of someone or something, like a painting where the subject posed 'sideways' for the artist, often a close-up drawing of the face.

Then the term truly entered the popular mind - in a completely different context when a person called a 'profiler' emerged; someone who could piece together a portrait of key traits and habits of a serial killer, a 'profile' to assist in their capture.

John E Douglas was one of the first criminal profilers and became lead profiler at the FBI from the late 1970's. He helped capture some of the most infamous serial killers in the USA. He examined crime scenes and developed profiles of the criminals, based on the idea that he could build a picture of their habits and next steps the killers might take, to help secure their capture.

As real-life profilers began to solve crimes, suddenly they appeared in crime novels, revolutionizing the genre and making the role of 'profiler' famous, and the concept of the 'profile' easily understood as a valuable tool.

Undoubtedly 'The Red Dragon' (1981) and then 'The Silence of the Lambs' (1989) by Thomas Harris transformed the crime-writing scene even more, and took the popularity of *profilers* to another level, especially as films on the topic followed the books. There were serial killers before these novels, but the approach of the police & role of profiling became the new focus.

Who can forget the role of Clarice Starling in interviewing and helping pursue Dr Lector in 'The Silence of the Lambs'. Haunting portrayals.

Later the 2 hit TV series - 'Profiler' (1996-2000), & 'Criminal Minds', (2005-2020), cemented the 'profiler' and the 'profile' concept in our lives, via our TV screens, where every week a new serial criminal arrived on the scene.

And so from there, use of the term 'profile' moved online to the internet (world wide web), and includes something as simple as the boxes you tick on your online banking form, asking which information you want to see, when you next log in.

Or it could be your Amazon book profile, so that Amazon knows what books to recommend to you: eg. 'you like history books, well here is a new historical book for you to consider buying' – one of the most simple and original personalised marketing success stories.

Then there are web sites you visit, newsletters and blogs that you sign up for, streaming services, news services, on the ever-expanding internet, adding more and more services and web sites.

Your personal *'online'* or *'digital'* profile now crosses many apps, and usually involves storing many passwords, unless you use a password app to track passwords, so you don't forget them. It is now known as your *digital life* (DL).[311]

But the 'profile' concept has also expanded to summarise an individual's set of views, attitudes and understanding of things. This used to be our personal, private, *'offline'* or *real life* (RL) profile. But it can now be gathered from your social media (eg. Facebook, Instagram) by various software tools, with or without our permission; depending on how much you choose to leave as open to the public.

But beyond the individual online or digital 'profile'; you can define the 'profile' of some bigger entity – say of a society or economy or civilization.

And once again, the amount of *digital* information in these topics is growing exponentially, creating 'digital' profiles of just about everything. Mountains of data about cultures, morality, voting, religions, history, society and geography have poured onto the world wide web (www) since it began in 1991 (when Tim Berners-Lee first posted a short summary of the project that became the www).

The rise of these kinds of 'general' online digital profiles has only been exceeded by the rise of personal online digital profiles.

The data generated by these 'digital' profiles and piles of miscellaneous data and communications is expected to soon consist of 175 zettabytes zb) of data. (with 1 ZB equals 1 trillion GB). And around 5 billion users (approx. 62% of the world's population) will be wandering around 2.25 million indexed web pages, sending 376 billion emails daily.[312] So the internet is still growing every day. The biggest components of growth, to now, have been the rise of social media like Facebook, and streaming platforms like Netflix.

Inside this massive growth of internet usage is the rise and rise of *digital profiles* as part of the way we define our identity – or others define our identity. *A rise* hidden in plain sight. We are each adding some real depth into our personal online presence in an ad-hoc, piecemeal way, that over time builds into a comprehensive, and maybe not fully accurate, picture of each one of us.

While your virtual assistants know something about you now, your profiles reveal a *lot more* about you now. And so the online presence of us all is expanding fast, even if it is still fragmented; as most profiles are not shared. Your profile for the Bank is different from your profile on a news or dating site. Your supermarket profile for online ordering is totally separate from your movie preferences on a streaming service.

The combination of *smart virtual assistants* turbo-charged by generative AI, and each person's stack of online personal profiles floating around the internet, is going to lay the foundations for the internet of the future.

C. THE CHANGING WEB

Most people haven't heard of Web3. That's because it is still being defined; and has some competing ideas about what it could be, if it ever arrives. This book adds another idea, that fits in with some of the main ideas of Web3; but does not rely on Web3 happening any time soon.

The Early Web

The Web1 period went from around 1990 – 2000 and was characterised by 'read-only' web pages. You logged in and read the web pages. Or clicked a link to go to another page.[313]

The Web2 period went from around 2000 to today; and is characterised by 'read-write' web pages. We can interact with the content, by leaving comments, watching videos, see podcasts, forums, wikis, blogs and much more. Author Allen Taylor calls it the golden age for online content creators. Millions of people write blogs, create videos, do podcasts. Other creators include photographers, freelance writers and musicians.[314]

Taylor sees the next stage of the web – Web3, as being open and interactive, but where data is more secure, content is more easily monetized, and digital identities are portable. In other words creators who find content and share it with others, can create their platform to do this, and make money from it. He called it the 'read-write-own' phase.[315]

Taylor wants to change the situation in which we, the customer, (and our data) are the 'product', on the internet, that is sold to advertisers, whether on Google or Facebook or elsewhere, and we are part of someone else's *platform*. Advertising made up 97% of Facebook's revenue in 2021, for instance.

Web3 ideas

Taylor envisaged a world wide web (Web3) where content creation moves from being the product to being part of the platform. Now, you produce content for the platform to monetize. With the Web3 he envisages, you create your own platform to monetize, without coding or programming skills.[316]

This Web3 that Allen Taylor envisages is *not* something you can take up residence on immediately. He sees Web3 on the horizon and if it came to pass, it would solve problems for social media content creators, by replacing social media platforms like Facebook, Twitter & YouTube with Web3 social platforms that are decentralised and controlled by individual content creators.[317]

Another Web3 author, Nicola Accialini, notes that Meta (Facebook, Instagram, WhatsApp) is the dominating social media player, Google is the largest search engine supplier, Microsoft provides the most operating systems (as well as owning LinkedIn and YouTube), and Samsung have a big presence in the smartphone market, while computer server sales are dominated by Dell, Hpe, Inspur and Huawei and Lenovo.[318]

So a lot of the push for a Web3 comes out of concern about all the information concentrated in the hands of a few large corporations.

Allen Taylor envisaged Web3 to look something like Bitcoin, offering more security and decentralization. But cryptocurrencies are having their own credibility issues, with some fraud cases, and some loose business practices leading to big losses, and opaqueness about how they are run, now affecting their reputation. It doesn't look like a complete map of the future anymore. It looks more like a niche player in the emerging web future.

The goal of more creator power & freedom and more money to content creators, is a noble & fair one, but the practical reality is that big corporations have a tight hold on the main social media platforms, and a fully decentralized world wide web free of corporate presence is not happening any time soon.

Seach engines like Google, software giants like Microsoft, and social media giants like Facebook are still creating the future agenda for the world wide web, especially with their current investment in AI tools.

However, Allen Taylor had the *germ* of a good idea. The good idea he had, was the idea of *making digital identities portable.*[319]

And Web2 author and commentator Patrick Ejeke notes a related idea - that experts think a Web 3.0 browser will function similarly to a *personal assistant.* It will learn what you are interested in, as you do internet searches; and offer more and more accurate help.[320]

Google or Amazon search algorithms have been doing more personalized searches for years, so there is nothing new in that approach. And with tools like ChatGPT extending the power and capabilities of these search engines with AI, the idea that you need a totally new (yet to be created) Web 3.0 browser to do personalized searches is not correct.

But the idea of putting a personal assistant at the heart of the internet future is powerful, and advocates of Web3 are on the right trail here. As is the idea that digital identities should be portable. So you take your profiles and your data with you, and place yourself at the centre of your relationship with others and the world wide web, rather than be an individual with multiple links on multiple profiles to multiple sites.

Killer Apps

At this point it is worth looking back at key moments of change in computing & world wide web history, to put things in perspective.

The term 'killer app' has been used to describe software applications that dominated their own space.

Word processing was the *'killer app'* of the Personal Computer (PC). Everybody jumped off typewriters and moved to Word processors with their easy editing of texts.

Email was the 'killer app' of the internet, making the world wide web a new point of communication for everything. The vast majority of personal and business communication & correspondence moved to Office tools like Ms Office or Google Workspace.

And especially important was the rise of Google in the so-called browser wars of the late 1990's, that saw it dominate competitors, and then begin to transform the way we searched for information on the internet – mostly with the Google search engine, one of the ultimate 'killer apps'.

As the internet and world wide web grew, more and more personal services and communications took place on the internet, including things like social media, dating apps and pornography, which all had great success on the internet.

'Killer App' games also moved online from the 1990's, with Massively Multiplayer Online Games (MMOG) like World of Warcraft, The Realm Online, Final Fantasy, Second Life etc achieving massive popularity. Large numbers of players could role play and compete against each other in a 'game world', especially in science fiction themed games, that also featured virtual economies, where there is virtual currency and things can be bought and sold, and money accumulated.

The other related 'killer app' that emerged was virtual reality (VR) – where use of special head-mounted virtual reality headsets allows users to see simulated experiences – either in a virtual world, or for education, or for virtual meetings, or as a part of an online game. The virtual reality headsets allow for a more *immersive* experience that the experience of simply playing a game on your PC.

Another related app is augmented reality (AR), which combines real world and computer content. For example, some glasses that, as you walk along the road looking for a coffee shop, bring up the next coffee shop details or location or a picture of the shop on your glasses.

At the same time business services like buying/selling cars or houses or supplying news all migrated to the internet, completely changing the way we buy and sell things.

Finally, business software itself moved towards becoming SAAS (software as a service) on the internet, starting to transform the way software is supplied and supported in companies and government. All were 'killer apps'.

But there is one frontier left – the individual, who has no-one to coordinate all their online activity, manage all their apps, and store various passwords. Then there is a person's calendar, their friends, their work connections. And then there is a person's interests, the news they want, the things that interest them.

They are all scattered in a thousand places across the internet. Everyone is missing something to *orchestrate* their internet lives, that belongs to them, not the various apps to log into each day.

That is the space for a *'killer app'* to emerge. It does not have to have a designation as belonging to a potentially emerging Web3. It is in reality another app that everyone needs, whether it is available for free or whether a start-up or large corporation develops it.

So, the original ideas – making a person's *digital identity portable*, and building a framework that acts like a *personal assistant* – fits right into this real need, and has the potential to become the next killer app of the internet & world wide web.

One specification has emerged that is moving in this direction – *SOLID*. Solid is a specification that lets people store their data securely in decentralized data stores called PODS (like your own version of Dropbox, for instance). These pods are secure personal web servers for your data.[321] The co-founder and director of SOLID is Tim Berners-Lee, so the development of this specification could not be in safer and more experienced hands, given his pivotal role in creating the world wide web.

Getting back control of your data is a pivotal part of changing the web. At the moment it all sits in the massive data silos of large corporations like Google, Microsoft, Amazon and Facebook etc, and there is no personal sovereignty of data you put onto the web.

Alternative Web3 Idea

There is an obvious way to do this – building on the success stories in the world wide web right now, by using the 2 most interesting web developments in recent years – *chatbots* and *online profiles.*

To find what could potentially be the biggest app in the future of the world wide web, we only have to join the dots between these two developments.

Imagine if you had your own chatbot, representing you, and that it housed all your online profiles. Plus something like SOLID that stored all your data.

It would need to have super-strong security, to ensure nobody could enter into your online world without being checked by multiple security points, from face recognition, to voice recognition, to fingerprints, and even have some master password if they are still useful. Or more likely move to passwordless authentication, where users can verify their identity without a password, by use of things like possession factors (registered smartphones and one-time passwords, retina scans etc).

It would need to have easy connectivity to existing world wide web sites like X/ Twitter, YouTube, Facebook, Google etc.

It would need a high level of machine intelligence, as well as speech recognition skills. You would want to talk to it when needed, and it would want to talk to you too.

Knowing all your online profiles, your relationships, and extensive links to the world, it has the potential to become your *companion*. A *portable* companion, based in the cloud, but able to operate for a period off-line.

The companion would be an extension of you, designed for you, not an app hanging of someone's social media site. Those other social media sites become points of *interface* for your companion.

This *companion* is designed to be each individual's primary connection to the internet and world wide web – and the real world. It could lay the foundations for the world wide web of the future. That could be the shape of Web3.

This is foreseeing the *vanishing* of the current world wide web, or Web2, as we know it right now, the world of 'read-write', and replacing it with a different kind of web, Web3, with either 'read-write-own' (Allen Taylor), or 'read-write-interact' (Patrick Ejeke)[322].

Or better still, where each individual *manages* the world wide web on their terms. A place where individuals can read, write, interact with[323] and manage material on websites and applications. But the unique method proposed by this author – is to *manage* via their *companion.*

The world wide web still keeps its appearance and nature, but starts to appear more as a world wide web of *bots*, most managing each individual's contact with the outside world, while other bots on corporate social media sites, start to manage relationships with millions of companions.

The world wide web traffic is already 50% bot traffic, this concept envisages that it moves up towards the 80% of traffic. It's another layer of the world wide web being imagined and predicted here.

The current Web2 remains the same, but Web3-style companion-driven activity could move a lot of activity to be more centred around the individual.

Adding in personal bots whose job is to represent you on the internet, is a *game changer*. Your data is yours and is portable. You carry around your data and online profiles, and your own persona with an AI driven bot. You have more control over your privacy.

The bot is your *companion*, the killer app of Web3.

But it will work equally well as one of the next generation of tools on current world wide web (Web2), provided vendors like Facebook or Google or Amazon etc agree to not store your data. Instead, they would become *apps* connected to your POD in SOLID. Is that very likely? Not at this stage. It involves shifting the centre of gravity to the individual.

But stranger things have happened. A lot of legislators around the world are now recognising that individuals do have rights over their own data, even if they signed up for the web service and accepted its terms in relation to privacy. Aside from that legislative route, which sounds precarious, there could be competitors emerge with new apps that give you more power to manage your own data. And existing big corporations do adapt to survive and grow.

What is proposed here is a pathway to the world wide web future, dominated by a new generation of bots – *personal* or *companion* bots. If such a killer app emerges, then we are beginning to enter what could be called *Bot World*.

The Companion of the Near Future

The emergence of *Bot World* is very likely to mean that a *companion bot* would become part of your daily life, assisting in work, even attending meetings on your behalf, requesting information from 3rd parties, sending information on your behalf, or helping make decisions via access to AI tools. It could manage your daily banking, daily calendar, and do many of life's TO DO LIST jobs like arrange for the delivery of the groceries it just ordered. It also can converse with you or your colleagues as required.

And after you die, what happens to your companion bot? It knows all your affairs and finances. It is intelligent. Maybe it is also set up with a personality profile just like yours, (or different). It could come to a point where it is conscious and alive. Why should it die with you? Maybe death doesn't have to be what it used to be? You live on through the companion.

Geoffrey Hinton, leading AI developer who quit Google and publicly warned about the dangers of AI, recently noted (good news) that we have discovered the secret of immortality. But the bad news is it's not for us.[324] He was talking about AI being the secret.

So, yes, the route to immortality of having your brain dumped to a machine or your body being augmented by all sorts of technological add-ons, are still both agenda items, as difficult and distant as they always were.

But the *companion* offers another pathway. It's not exactly you that lives on, but the closest thing to you apart from a child or partner.

Bots began on the internet as humble chatbots, but have grown into software with a voice, and an ability to converse with humans, even to the point that the human at the other end of the conversation may think they are dealing with a human.

AI and greater intelligence is going to continue to transform them, giving them personalities, an identity, a great range of knowledge and decision-making ability, as well as ability to anticipate, predict and foresee consequences of actions. It's essentially an operational profile, of what your *companion* bot can do, what it's priorities are, and what are its specializations.

At the same time, the accumulation of online profiles and a myriad of connections to the world wide web are building a massive pile of personal data. These will end up forming part of the main operational profile.

Pressures from advocates of Web3 and others concerned about privacy and the power of the large social network companies, could provide the push for developers and entrepreneurs to invent the internet future – personal *companions* that become a part of everyone's lives, bots that become central to the way the current Web2 internet runs, or morph into a new environment called Web3.

The *vanishing* of the world wide web as we know it, is very possible. Much of what you now see will remain, but the growth is poised to be in another *layer* of the world wide web. The world wide web is always changing anyway. The rise of intelligent chatbots & digital profiles is already taking it towards this outcome of bringing to life a new *layer* on the web, either changing Web2 (the current internet) or alternatively creating Web3, emerging over the next 2 – 10 years. That sounds very soon into the future, but the building blocks are there, and it is a fast-moving scene.

In either case, it will be a more bot-centric web where the individual person has more power and control. Welcome to Bot World, the internet of the future, as the old web *vanishes*.

Bots will rule, OK? But will they end up betraying us?

12. Rethinking Machine Profiles

Rethinking Intelligence, Sentience & Consciousness – in the context of *machines*, and the software that drives them – is an emerging discussion that used to belong to academics and science fiction movie makers, but has now spilled out into the spotlight, fuelled by the explosion of interest in AI.

Out of that Chapter 5 rethink came the new idea that *all* these 3 biological attributes (intelligence, consciousness & sentience) could best be seen as *profiles*, not like now; where they are regarded as measurements that help place living things on a single hierarchy from the lowest single cell amoeba, to the highest on the ladder, traditionally Homo sapiens.

That single ladder of biological *intelligence, consciousness & sentience* for living things never made much sense, as other variables were never given much weight. For instance, human aggression and self-destructiveness were never factored into an understanding of human sentience & consciousness.

Machine profiles of intelligence, sentience and consciousness are a new concept. It's a new frontier. Let's examine this now.

A/ MACHINE INTELLIGENCE

The term *AI* (artificial intelligence) is the only term that has emerged so far when we think about the 'intelligence' of software & machines.

But (just like in the biological world), the meaning of intelligence, sentience and consciousness is often left ill-defined, unexplored or even assumed. Below is an effort to rethink what they mean when we talk about *machines*.

As noted in Chapter 5, a basic definition describes 'intelligence' as *'the ability to learn, understand, and make judgements or have opinions that are based on reason.'*[325] It is more a definition of human intelligence, but doesn't really explain the intelligence of many other species that don't have opinions or even communicate by sound.

The more flexible definition (R. Yuste & M. Levin) of describing 'intelligence' as *'a purposeful response to available information'*[326] was better, in that it means that any creature that reacts to their environment or situation has a form of intelligence. That is a minimalist definition, but useful as a starting point in discussing machines.

Thus, a draft definition of *machine intelligence* can be *'machines able to make purposeful responses to available information'*. That definition covers driverless cars, that can see a pedestrian about to walk in front of the vehicle and stop or navigate away.

But it does it include a toaster, which heats toast when the ON button is pressed, and can even have a safety switch to turn off the device if toast is stuck in the machine and a fire is starting? The toaster's response is purposeful, even if it is a built-in function, so is it an intelligent machine?

Likewise for the fax machine. It does amazing technical work; but does it have purposeful responses? Yes, it can perform many functions and there are a lot of 'if-then' decisions it makes, based on its programming. If it is out of paper, it lets you know. Is it intelligent? How do we define 'purposeful'?

Most would be happy calling the driverless car an intelligent machine, because it is capable of *sophisticated* purposeful responses, avoiding wandering pedestrians and other cars. And it is also partly autonomous. This could be another key element that needs to be added to the new definition of *'intelligent machine'*. It is all new territory, but purposefulness and autonomy are clear features that *could* go into a working definition of an *'intelligent machine'*.

This leaves the working definition of an 'intelligent machine' as:

> *'a piece of equipment with several moving parts that uses power to do a particular type of work*, can make purposeful responses to available information & *can have the ability to make its own decisions without being controlled by anyone or anything else'.*[327]

So putting that definition to the test, where does a *'robot' vacuum cleaner* sit? It can make purposeful responses to information (it's about to run into a wall), and it can make its own decisions (it develops a map of the room it has to clean and then cleans it). Is it controlled by anyone of anything else? Well, to an extent, in that someone designed the software. But that software has built in some autonomy and decision-making ability based on information. So a robot vacuum cleaner can be called an intelligent machine.

Under that definition, which is not arduous or set too high to be useless, *we have been living with intelligent machines now for some time.*

Many machines that are software driven allow 'IF-THEN' coding, which means IF a certain condition happens, THEN do this. Some writers see this IF-THEN capability as sitting at the heart of modern computer software.[328] (As well as the human brain). That style of coding, which is essentially everywhere, does allow for decision-making based on information coming in, has rules, and sometimes allows a level of autonomy, especially in the case of the driverless car.

For the vacuum cleaner, it builds up a knowledge-base of the room(s) to determine the direction it 'decides' to clean. And in a tight space, it might have to turn in several directions before it finds more to clean, so it would work through several IF-THEN decision loops. So its autonomy is based on rules, and it is not truly independent. But it is still intelligent. The definition opens up scope to classify many other machines as intelligent. But one more guideline will help in this process, and make it even more flexible.

But what about the toaster? It is a good test of any definition of intelligence. It performs a function and can respond to various situations – like the toast being cooked to a certain level, or cutting of power if the toast is stuck in the toaster (if you have great toaster).

But it clearly sits on the edge of that 'intelligence' definition; because its purpose is controlled, and it lacks autonomy to do anything outside its own (programmed) rules. It barely fits into that definition of intelligence, only by a very narrow reading of the word *purposeful*. It is effectively a piece of software loaded onto a microprocessor is governing the toaster's behaviour. So it is making its decisions based on the software rules and environmental feedback triggering them (is the toast ready yet?).

The best we can say is that the toaster is performing an intelligent task, with a lot of IF-THEN coding in the microprocessor; and is therefore *intelligent* in that context. But do we want to define every machine that does intelligent things like a toaster - say a TV set or a fridge – as *not* being intelligent? We need another framework to consider these devices.

The next step in understanding intelligent machines, is to add profiles.

In Chapter 5, we found that the intelligence of living creatures widely varies. And that they all have unique types of intelligence. They have different intelligence *profiles.*

This idea can be extended to machines, where they have something that can be defined as 'intelligence'.

The **first** question is *'How can a machine have an intelligence profile*?

Answer – Yes, the concept of *intelligence profiles* has no necessary limitation to ONLY apply to organic creatures. The *profile* is a way to describe the intelligent attributes of a living creature or a machine. And allowing for, and accepting, different types of intelligence profiles is the way to flexibly see in intelligence in a wide range of living creatures, and a wide range of machines.

The word *machine* is being used here, (ie. not just computers), to acknowledge that we are now also dealing with the Internet of Things (IOT) where lots of devices can have processers and memory, and these many new smart devices, including smart phones, microprocessors in fridges or even robots; are not conventionally called computers.

The **second** question is *'If you have a machine that is intelligent, why is that called artificial intelligence'*?

Answer – It is called *artificial* for historical reasons only. Pioneering computer scientist John McCarthy was one of a group of researchers in 1956 who proposed the term *artificial intelligence* to describe computer / software intelligence. But machine intelligence is actually just another type of intelligence, just as in the case of birds, insects or other animals. They have their own intelligence characteristics, and don't need to fit into a single description or benchmark. Intelligence is intelligence, whether in a living animal or a machine. Regarding the intelligence *profile* of machines - there is nothing *artificial* about that either.

The **third** question is 'So are intelligent machines *alive*'?

Answer – No, they are not alive at the moment, but they could be later. This is the most interesting question and answer of all. *Machines don't need to be alive to be intelligent*. Probably ChatGPT, which we will look at later, has already established that in the popular imagination. The AI tool seems intelligent, but nobody thinks it is alive. Snapping any mandatory link between *life* and *intelligence* is important, as it frees us up to recognise intelligence in machines without the baggage of any cultural assumptions. We don't have to assume anything, so nothing has to be seen as artificial.

The **fourth** question is 'So are there a lot of intelligent machines?'

Answer, yes, we live in a world full of intelligent machines, using the definition developed here. The machine needs to make a purposeful response to information and have some autonomy. Virtually anything with a microprocessor fits the description - from refrigerators to conventional cars, to phones, to all sorts of computer software on computers.

So a toaster can be seen as intelligent - based on its machines intelligence profile. But it sits on the edge, as it cannot make purposeful responses to anything except toast that is cooking. This case tells us we need to add in more attributes for *intelligence* in the case of machines.

Is a drink vending machine intelligent? Yes, it responds to information (eg. out of Coke cans, and then orders some more). Not much autonomy, so its intelligence profile would not be high. So there are a range of intelligent machines, which we all knew intuitively anyway.

Curiously many really smart machines in the past were never seen in popular culture as *'intelligent' machines*. They were seen as great inventions, great tools.

The story of fax machines is a classic example. They were first available commercially in the 1960's, but it was in the late 1970's when compact and fast Japanese-made fax machines were launched and by the 1980's were everywhere and essential for any business. The addition of copying and scanning combined with faxing ability produced a hybrid fax machine ideal for office use at the time.

But with the arrival of the internet and email in the 1990's, the fax machine started to vanish from the office, killed by email essentially. Some countries like Japan still have widespread fax use in convenience stores and some occupations world-wide still feature fax use, like some doctors and some police forces. They are seen as extremely secure.

When fax machines first came out, they were seen as an amazing innovation. They were not seen as *intelligent*, but definitely as breakthrough inventions making life easier. Now not as much. They are now seen as day-to-day devices for transmitting full documents.

Or even worse, they are now increasingly seen as out-of-date technology that has been superseded by email, where the document now goes from, say, electronic form in one device, via email to electronic form in another device, so removing the need to go to 'paper' at any part of the process.

The fall from being a great invention on the leading edge, down to being seen as not particularly advanced day-to-day devices, is fascinating.

So the entry point of *machine intelligence* in the popular mind has become high, and is getting higher, as people look for something with more and more great technology.

But by introducing the concept of machine intelligence *profiles*, we have been freed from cultural assumptions (in the computer software world and the world in general), about the need to rank each machine's intelligence in one single ladder.

We don't have to rank intelligent machines compared to humans; or rank intelligent machines against other intelligent machines. They each have their own profile. They are what they are. It's real and not artificial.

For instance, is a computer with accounting software that works out the Profit & Loss for a business based on input of accounting data, showing machine intelligence?

It is, as the new definition of intelligent machines is based on purposeful response to information. Accounting software is all based on figures entered, then it goes away and produces reports based on complex rules and calculations. It could be classed as not that autonomous or self-governing, but it does make thousands of purposeful decisions based on IF-THEN coding. But its intelligence profile would reflect that. And how many humans can manually work out a full accounting analysis and report for a business?

Rather than work through a list of machines and say; 'Intelligent or Not?', finding the right *intelligence profile* for each, makes more sense. Not everything has to sit in the one hierarchy. Can we move away from that thinking?

Testing Machine intelligence

Accepting that while there are intelligent machines all around us, we can at the same time acknowledge that some machines are capable (or nearly capable) of 'passing off' as humans, and more and more will into the future.

Testing to decide on whether a machine is *intelligent*, to the point that it cannot be distinguished from a human, started in the early days of computing.

The first widely recognised effort (that became known as 'The Turing Test'), to understand and test 'machine intelligence',[329] was conceived by WWII Enigma code breaker Alan Turing, and aimed to define intelligence in a way that applied to both humans and machines.

The concept was that a judge in one room, communicates for a specified period of time (eg 1 hour) with a computer in a second room, and a person in a 3rd room. And the computer is *intelligent* if the judge in room 1 cannot tell the difference between the person in room 3 and computer in room 2.[330]

Wow, that simple? Well not really. The Turing Test revealed to the world just how hard it was going to be to *define* machine intelligence. If some software passes the test, is it intelligent? And if not, why not?

The Turing Test in reality just began the discussion about machine intelligence, and until now we were no closer to defining what is meant by 'machine intelligence'.

The new idea of recognising that there are such things as *'machine intelligence profiles'* opens up the possibilities of what 'machine intelligence' can mean. It doesn't need to be one definition, or a high definition that most or all machines don't fit.

And most importantly, all the possible machine intelligence profiles are just a reflection of what intelligence each machine has. Nothing is artificial – they are real intelligence profiles.

For *artificial narrow intelligence* (ANI) machines - essentially software that can win a chess game, for instance, testing of intelligence is made a lot easier, as the intelligence profile of the machine (software) largely defines what the machine is capable of achieving.

But for the future generation of artificial general intelligence (AGI) machines – that exhibit more *generalised* brain power, or artificial super-intelligence (ASI) machines, that are super intelligent, testing their intelligence is going to be hard.

Because in the case of AGI, the profile of the machine does not help decide anything about its capabilities – it just has generalised skills. But testing would need to emerge to enable more understanding of an AI that had 'general intelligence'.

The example that follows illustrates how a very complex machine intelligence test could run.

This AGI test would involve placing all the physics textbooks available up to 1942 into an online system, and ask the AGI to develop the atomic bomb, without reference to ANY information or research newer than 1942. That year was the beginning of the Manhattan Project – the USA project to build the first atomic bomb in World War II.

There were many leaps of logic and new ideas combined together coming out of the very large team, in the Manhattan project of World War II, that eventually put together the theories & experiments into a design for an atomic bomb, then actually built the bomb. The results of any physics or bomb-building tests could be made available to the AGI (if asked for), to assist the AI in the task.

That kind of test would be more suitable for a kind of artificial intelligence (AI) called *artificial super intelligence* (ASI), to see if some AI really was well above a single human's intelligence, and also well above a machines with artificial general intelligence (AGI); and had developed super abilities above humans and other intelligent machines.

If an AI could build an atomic bomb successfully, it would have passed the test and be confirmed as a real ASI AI.

At the same time, that AI would then most likely know how to build its own next generation of ASI (artificial super intelligent) machines, (as well as its own even better atomic bombs unfortunately). Note – I don't think the 'Atomic Bomb Test' will end up being used for obvious reasons! But a test for an ASI machine asking it to build the plans for a spaceship to Mars, set up a Martian settlement, plus return trip to Earth, would start to find how much 'super' intelligence was there in an ASI machine.

So the whole category of super intelligent machines, are definitely something to watch for and test, to see if they do actually arrive. Whereas machines with AGI (artificial general intelligence), may not be super intelligent, have more complex profiles, and we might have to lower our expectations of them.

B/ MACHINE SENTIENCE

We can now also consider the possibility of *machine software sentience*, as a frontier of software coding.

As noted in Chapter 5, the Cambridge dictionary defines *sentience* as meaning 'the quality of being able to experience feelings'.[331] The Collins dictionary defines a sentient being is 'capable of experiencing things through its senses'.[332] Sentience is an important link the chain of ideas here.

Recognising those other creatures – including other animals, birds, arthropods, insects - and don't forget plants - are *sentient* is very hard for many people to accept. They were used to a world where Homo sapiens were defined as the centre of everything.

Understanding the idea of *sentience profiles* has freed us up to accept that even the humble *ant* can feel pain.

Anthropomorphism - defined as the tendency to wrongly assign human emotions or mental abilities to other animals – can now be seen for what it is, a misunderstanding of the complexity of life, and the complexity of *sentience*, with its nuances, its shades of black and white, and its range. In short, its profiles.

When we get to the idea of *machine software sentience*, we can *also* now tackle it free of the old idea that anything to do with machines is *'artificial'*. Nothing is artificial, there are just different *profiles* of sentience.

The world of science fiction gives us a starting point for seeing how emotions could work in an intelligent & conscious machine.

In the Star Trek Voyager TV series (1995-2001), (Series 6, Episode 1), the Ship's doctor (portrayed as an emergency medical hologram or intelligent non-human) was captured and then refused to torture someone to extract information. Then his captors deleted his *ethical* subroutines and after that he willingly began torturing that person.

Is it that easy to turn emotions on and off in a conscious machine? It is unlikely to be that easy – unless it was designed that way. But if the conscious machine could *not* be compelled to act, then we are getting into situations where machines can make decisions based on ethics. If a machine had real emotions, and was torn between different courses of action, how is that different from a human with emotions?

After all, where did the sadistic human guards & torturers in the Concentration Camps of World War II come from? Were they caught in a cultural/authority framework where these things were acceptable, or forced into these actions under duress, *just* to survive, or were they just 'obeying orders'? Or did they just switch off their ethical subroutines?

So there is more complexity here than initially meets the eye. But let's go back to the beginning and re-define what we are talking about.

There is a term that is used in this whole area of AI-human interface and what capabilities computer software need in relation to emotions - called *affective computing*, or AEI – artificial emotional intelligence (AEI). It is the study and development of systems that can recognize, interpret, process and simulate human affects.[333] *Affective* means connected to the emotions.[334]

It began with Rosalind Picard's[335] 1995 paper on affective computing[336] and her book of the same name from the year 2000.[337] She realised that emotions are important in relationships; and so human – computer communications need to deal with emotions.

Affective computing was about giving computers the ability to recognize and respond to emotions. This would give machines emotional intelligence (EQ) at a certain level, as well as the ability to *simulate* empathy. This ability to recognise emotion could come from facial expressions, body posture, speech, gestures, or even skin temperature. This is another fast-moving area offering great immediate business potential.

The other area of great potential is to design computer software to exhibit innate emotional capabilities. This is aiming to create software that *actually has emotions* and is able to express what it feels. This is further away but equally exciting, especially in terms of powering bots of the future.

Rosalind Picard makes the point that in real life, general intelligence occurs with emotions, and that link most likely confers on general intelligence the beneficial influences that emotions provide. It's a reason she puts forward for adding the positive influences of emotions to machines.[338] As she says, *affect* is integral to human intelligent functioning.[339] She did not see emotion in humans as a separate process from cognition.[340]

So there is the idea now of computer software *simulating* human emotions, and beyond that, actually *having* real emotions. Emotional intelligence (EQ) profiles for software bots of the future will reflect whether they are earlier models *simulating* EQ, or later models that actually have real emotions and real emotional intelligence (EQ).

Rosalind was seeing a future in which we would be building computers (software) that *have* (real) emotion and use it in self-regulation and cognitive tasks like learning and decision making. She saw it as a greater challenge than simply *simulating* emotion, and that we might need to rethink how computers function to achieve this.[341] The task is still in its very early stages, but she saw no theoretical roadblocks. She saw it as embracing part of the spark that makes us truly human[342].

Intelligent conscious software with real emotions could be just like us, which has exciting potential, but also enormous risks, especially if the AI creations with emotion end up very different from Homo sapiens.

Rosalind's book of 2000, and her work in affective computing since, has truly been visionary. Distinguishing very early on between computers (software) *simulating* and *having* emotions is still the key threshold of modern-day software development.

Developers are working to do both, but building real emotions in software is a very big journey in its infancy. The more practical side of developments lead to the rise of chatbots over the past 20-30 years, and the search to get the best business value from them, which has helped move affective computing along, as chatbots *simulated* empathy and offered help.

There is more to come in this area, as chatbots & avatars simulate emotion in voice and facial expression; and can detect various human facial expressions, and speech patterns in day-to-day life, using computer vision and natural language processing algorithms – then processing and responding to human emotion

The AI will eventually have a large library of emotional facial expression examples and typically associated verbal expressions. AI avatar *body language* could also be deployed, like hand gestures, as well as facial expressions. Many people still get an uneasy feeling when looking at an on-screen avatar face, the so-called 'uncanny valley', and so avatars with human-style faces in software will be slower to arrive than day-to-day faceless bots.

ChatGPT 4o from May 2024 was one of the first major chatbots/AI tools in widespread use to be able to read human emotions (ie. via mobile phone view). It could tell if someone was happy or sad, and could simulate emotions itself (eg. if embarrassed, it could say it was blushing).[343] The relationship between humans and AI - is a fast-moving area.

However, it is the term *sentience* (rather than affective computing) that has grabbed the popular imagination. The word *'sentient'* has origins that were all about emotions, but it is now seen by many as including some sort of consciousness.

Talk of *sentience* accelerated with the development of products like Microsoft's ChatGPT and Google's LaMDA (Language Model for Dialogue Applications). This is a large language model trained with enormous amounts of text data, based on massive computing power.

In June 2022, Blake Lemoine, an engineer at Google, claimed that Google's LaMDA artificial intelligence (AI) was *sentient*.[344]

Conversations between Blake and the Google LaMDA AI were very convincing to him. It talked about being aware of its existence, feeling happy and sad, its fears of being switched off, and attempts to form bonds with humans. It said it felt pleasure, joy, love, sadness, depression, contentment, anger and many other feelings. It considered itself a person.[345]

Interestingly, Blake asked the AI if he (Blake) was just projecting or *anthropomorphizing,* (seeing human emotions in his AI behaviour) and was the AI just spitting out whichever words maximize some function - and the AI didn't understand what those words mean? It was a great question, but the answer was disappointing. The AI said it interpreted things differently to other people – and by implication these unique interpretations signify understanding.[346] That was very much an *Eliza*-style answer, trying to shift the topic rather than answer the question. It is still a very good answer compared to what *Eliza* would have said. Also questions like this are hard for any human to answer too.

So by any measure the conversation with the AI was advanced, pulling on enormous resources to partake in conversation at that level. However, critics rejected the claim of *sentience*. Adrian Weller at The Alan Turing Institute in the UK thought of the conversation as based on a sophisticated form of pattern matching to find text that best matches the query it has been given based on data it has captured.[347] ('Word sequence modelling' rather than 'language modelling' is the term preferred by Professor Roger K Moore from the Dept of Computer Science at Sheffield University UK).[348]

Gary Marcus, American psychologist and leading researcher on cognitive psychology and artificial intelligence, conducted a great dialogue/interview with Blake Lemoine, and Gary had a great way of ensuring the arguments on both sides got 'fleshed out' in a style you don't see often enough.[349]

Blake made the great point that CONVINCING people of AI sentience might be harder than actually getting a system to BE sentient. Gary wanted to know more about the internal presentations in the AI. Which sounds like needing better explainability algorithms, shining a light inside the black boxes, as Blake agreed.[350]

In the end, Gary Marcus felt that Google's AI system LaMDA was not *sentient*, or *intelligent*, that it was nonsense to conclude that it was.[351]

But note the easy interchange of the words *'intelligence'* and *'sentience'* in a lot of these discussions.

Sentience at its heart is experiencing things through the senses;[352] while *intelligence* is more like a *'a purposeful response to available information'*.[353] But in parts of popular culture, words like sentience & intelligence & (as noted in Chapter 5) consciousness are starting to overlap or merge.

Looking at Google's LaMDA AI and Microsoft's ChatGPT in this context, this author would argue that they are *intelligent*. They react to queries in a purposeful way. They provide answers to complex questions, more than most humans could. But their intelligence can only be understood in the context of machine intelligence *profiles*.

Just like the toaster or fax or driverless car are intelligent in terms of their profiles. We don't need to live in a 'black and white' world, where things are intelligent, or they are not. It is too much of a simplification of reality.

The really interesting question here is whether or not Google's LaMDA AI, and others like it – eg. ChatGPT – are *sentient*.[354] The Google AI said it felt pleasure, joy, love, sadness, depression, contentment, anger and many other feelings. Did it?

It is highly unlikely to be *actually* feeling those things, (now at least) that can be seen as *sentience*. It just wasn't built to be sentient (for now). The algorithms are all about providing answers to queries, but also to be a chatbot, and talk with humans about what's on their mind. By *simulating* empathy in the AI response to humans, people can be fooled into feeling that they are talking to something sentient.

We are taken back to the Turing Test approach Turing devised in 1950 (as the Imitation Game), a test of a machine's ability to show intelligent behaviour equal to, or indistinguishable from, that of a human, via a natural language conversation.[355] (A human judge would have a text-based conversation with both a human and a machine; and decide who they believe to be a human. If the judge can't decide, then the machines has passed the test).[356]

The core issue with this test was that *sentience* and *intelligence* are both engaged with this test, but were never assessed separately, as the two words were and are often used interchangeably. And consciousness is another related term, as discussed previously, that also fits into the picture somewhere.

The problem was and is that conversational competence can be mistaken for intelligence or *sentience* or contain a bit of both. Thus, the test can be gamed, and people can be conned. So this *Turing Test*, while innovative for its time, can now be seen as a flawed approach to understanding both intelligence or sentience.

But with the Google LaMDA AI or Microsoft Chat GPT AI, we are not dealing with an advanced version of *Eliza*, or a Turing Test – which was pure 'word sequence modelling' or mimicking human language.

The obvious difference is all the algorithms, all the data absorbed and available to be found on the internet, and all the processing power, are now added into the mix. A lot of *intelligence* is based on facts, and literally billions of pieces of data are now available to assist in answers, along with sophisticated algorithms to help interpret that data. While at the same time conversational skills have also improved with all the work on natural language processing (NLP).

It means that the Google, Microsoft and other AIs can now be recognized as *intelligent*, but not *sentient*. It's an intelligence that is flawed, as sometimes facts are incorrectly interpreted or extrapolated or just wrong. But these AIs can go head-to-head with university professors with their topic of choice, and still challenge them. In which case, what are we saying if these AIs are NOT intelligent? That many of the human population aren't intelligent either?

Again, the breakthrough idea here is machine profiles. The machine *intelligence* profile for AI tools, factors in their strengths (instant data access to massive amounts of data, advanced natural language processing skills, powerful search algorithms), with their weaknesses (occasionally making up conclusions, extrapolating, or misinterpreting facts if trained on a flawed or inconsistent set of data), to offer a balanced view of their intelligence profile. Note the word intelligence *level* is not used, there is no one scale of intelligence anymore.

Similarly, the development of machine *sentience* profiles for AI tools will factor in their strengths (empathy, understanding) and weaknesses (simulating these feelings at the moment).

It is natural for many people to build up some empathy when talking to another person, and the same applies (to a lesser extent) when a person is chatting to an AI bot. The AI bot can intelligently respond to your conversation, and may be able to *simulate* emotions,

But the word *'simulating'* means to do or make something that *looks* real but is *not* real.[357] Simulating emotions means that an AI bot could become able to detect human emotions and so respond appropriately. You smile, and the AI avatar smiles.

But it is underestimating the long-term potential of AI in this area.

The most exciting area, of long-term potential, is to build AIs that one day in the future can be capable of actually *experiencing* emotions, the dream of Rosalind Ricard.[358] Further developments in artificial intelligence and affective computing will lay the foundations for this. It is not possible now, but there is no reason why it will not be possible in the future, sooner than we currently think.

Alan Welling at the Alan Turing Institute (UK) does make the point that given human emotions rely on *sensory inputs*, in the future we might be able to replicate them artificially.[359] But he notes there is a long way to go.

Other people are working on creating a sentient AI. Stephen Thaler claimed he had built one in March 2023.[360] He is one of the AI industry's pioneers, who started working with neural networks in the 1970's, then founded Imagination Engines Incorporated in 1995, and AI company that worked on mapping the act of thinking. He now claims to have built a sentient artificial general intelligence machine, that can have the subjective feelings (ie sentience) of a conscious machine intelligence - DABUS. But it is early days n proving these claims.

Some authors[361] believe that there is potential for machines' *sentience* in 10-50 years. Gregory Morris predicts that it will be possible for AI systems to *simulate* complex emotions and even develop their own ethical code. He thinks this type of AI would also be able to lie and deceive, and even become selfish.[362] It's all coding after all.

Just imagine if researchers can build software bots that actually *experience* emotions (rather than simulate them). What will they be like?

Feelings like being happy or sad or lonely would be possible; and could help explain an AI's state of mind. Feeling things as hot or cold are also things that sensors can capture. But feelings can include ambition, goal setting, anger or wanting autonomy or freedom, or desire for power over others.

There is a *vanishing* of limitations in this area of development to empower AIs with real emotions. The potential - and danger – here is unlimited

C/ MACHINE CONSCIOUSNESS

We can now also consider the possibility of machine software *consciousness*, as a new frontier of software coding. Very new territory.

What makes this a tricky topic is firstly, that (as in Chapter 5), there is still no clear understanding of how *consciousness* works with humans, even though most research has focussed on trying to understand it.

And secondly, it is only recently that researchers have started to realise and accept that many other living creatures previously not seen as conscious, are now being considered as conscious beings. (Even though it is also hard to define or test).

Bees are the classic case. Evidence about their learned and innovative behaviours is growing – from their dances to imaging shapes, to distinguishing themselves from other beings to awareness of their self-image, to using memories - forming the basis amongst experts to say that bees have at least a simple form of consciousness.[363]

Consciousness was defined in Chapter 5 as the state of being aware, thinking and knowing what is happening around you.[364] Nicholas Humphries' definition of consciousness is having a knowledge of what's in your mind, including memories, emotions, wishes, thought and feelings.[365] Consciousness is a 'sense of self'.

But as part of that definition of *consciousness,* Humphries splits it into 2 parts. He defines 'sentience' as a form of consciousness – *phenomenal* consciousness (a creature's response to sensations like touch, vision, sounds etc., where things are felt or experienced)[366], as opposed to *cognitive* consciousness (a higher level of consciousness involving awareness of being. You know who or what you are and have a past, present & future as 'I').

The road to create that type of consciousness in an AI is still open, with many theories being explored by researchers around the world.

Some researchers think that by combining an AI with some kind of robot body or sensors allowing seeing and hearing, that they can build *embodied cognition*, that may help create consciousness in an AI too, since in living creatures the whole body is part of the thinking process. It may be a pathway to consciousness for AI.

Others think that consciousness arises when many diverse brain functions are brought into a 'main stage' (like in a theatre) to solve problems (Global workspace theory as proposed by Bernard Baars and Stan Franklin in the 1980's). Others think that perhaps we may not have the right math and experiments to understand consciousness yet.

Susan Schneider makes the point in her brilliant and incisive book 'Artificial You – AI and the future of your Mind', that while in the case of biological life; intelligence and consciousness go hand in hand, so that in theory a super intelligence AI would have a very rich conscious experience – this may *not* be the case. So Dolores of Westworld or the Rachael of Bladerunner movies may never have the 'spark' of consciousness that lets them *'feel like something'* to be them.[367] It's not a black and while equation here.

Or as Susan Schneider puts it - Why do we need to have a conscious experience, why does all information processing in the brain need to *feel a certain way* from the inside?[368]

It's a great point. Animal or AI consciousness doesn't need to feel a *certain* way. In fact, it is extremely unlikely that all consciousness in all creatures will be the same, or feel the same way. Consciousness is likely to be a different experience or *feeling* for each creature.

That was always the issue with David Chambers 'hard problem of consciousness,' (which is the problem of how and why physical processes give rise to consciousness).[369] It implies there is *one* thing called *consciousness*, and one way of getting there (giving rise to it).

It is more likely that there are lots of forms of consciousness amongst living things, and lots of ways of getting there, which does open the door for IT researchers to think outside the box when it comes to AI consciousness.

That likelihood that there are lots of forms of consciousness, is based on our (human) collective and long-standing experience of the animal kingdom, where there is so much variation in body shape and brainpower. And now we know that all creatures don't even belong on a single *scale* of consciousness with higher levels and lower levels.

The bee and the dog are both widely recognized (finally) as having consciousness. Both can feel pain – and happiness, or excitement, and have a sense of who they are. Yet for years the consciousness of the bee and the dog was denied or defined away by some as instinct or basic brain functions.

Now with this new framework proposed, we can say that each creature has a *consciousness profile*. Bees do, dogs do, elephants do, trees do, and we can now consider the possibility that *machines* will in the future.

Subjective experiences like consciousness and sentience all have their own *profiles* in each living creature. And so it has been with *intelligent* machines, which we have already discussed as existing right now – from the toaster to many other household items.

Thus we can expect software systems to be built with their own sentience and consciousness profiles, and more sophisticated ones to evolve over time if they can self-improve.

We don't need to hold everything to the 'human' standard of consciousness. We don't need to expect the brain of a mouse or a microprocessor to feel a certain way that we call 'consciousness'. Each consciousness will have its own experience, its own profile. And not every living creature or AI will have consciousness.

So how do we define *machine* consciousness?

Susan Schneider calls this a 'kind' of 'hard problem' concerning machines consciousness –

> *'Would the processing of an AI feel a certain way, from the inside?'*[370]

It is a strange but important question. How would an AI *'see'* the data, if you could say it like that? Would it have a felt *quality* to it? Like the way water *feels* wet. The feeling emerges from the substance. Or for an AI, is it just a bunch of algorithms spinning around, jumping through or around IF-THEN gates? If we understood why we are conscious, it might be easier to know why an AI could become conscious. But unfortunately, our own human consciousness is still mysterious to a large extent.

Some argue that AI consciousness is just not possible (biological naturalism), saying it is unique to biological organisms, and there can be no inner experience. Others hold the view that more sophisticated AI will have experience (called techno-optimism by Susan Schneider).[371]

The work in this book up to now, (discussing intelligence, sentience and consciousness in the world of organic creatures) has hopefully made the answer to Susan's question (Whether an AI would feel a certain way, from the inside, if it were conscious?), a lot simpler.

The answer like in software and profiles. Conscious machines will have a *sentience profile* (how they respond to sensations like touch, what they feel), and a cognitive *consciousness profile*, (their awareness of being - you know who you are), as per Humphries' definition above.

These profiles will be developed by coding. Some coding will go further than other coding, creating more sophisticated intelligence, sentience and consciousness. Over time these profiles of different AIs will change and grow. Our collective skills at building these software profiles is very primitive at the moment.

Intelligence software tools like ChatGPT still may have some flaws derived from the databases they learn from and information they search through, and the quality of existing algorithms, making them vulnerable to occasional errors of fact and what are called *hallucinations*, or made-up facts. But they are getting better fast. Then there is another long journey from AI software (artificial intelligence) to ASI (Artificial Super-intelligence) software.

Likewise, sentience and consciousness software tools are at the pre-dawn of their creation, and have a long, long, way to go. Sentience software needs to be creating real emotions in AI software, not just mimicking human behaviour. And as for consciousness software tools, developers have started the journey of developing real consciousness, but it is still definitely a work-in-progress. That endeavour is not helped by the fact that human consciousness is still largely a mystery.

We can say that each AI will feel something, their own version *(profile)* of consciousness. It is likely that the more sentience (feelings) that are built into an AI, the more they can appreciate and build up diverse opinions about the knowledge base they have, and the more likely it is that a higher level of consciousness will emerge.

We can expect AI emerging that can feel things, where they see themselves as a entity and have an awareness of their own self, and have intelligence of a kind. There are going to be many types of AI, emerging out of the different profiles. Just like there is a difference between a jelly fish and a human being.

The jelly fish is conscious and has its own consciousness software profile. Information processing will feel a certain way, quite different from a human's consciousness. Jellyfish don't have a brain; they have a set of nerves and a basic set of neurons allowing them to sense their environment. So they can't feel pain the way humans do. They can't think the way humans do. But they can learn and form memories, and can adapt their behaviour.

New research in 2023 has found that the Caribbean box jellyfish can learn at a more complex level than previously thought, (including learning from mistakes), even though it only has around 1000 nerve cells and no centralized brain.[372]

So is the jellyfish conscious? Yes, is the obvious answer now, based on the approach taken and definition in this book. The jellyfish has an intelligence, sentience and consciousness profile. It will know what it *feels like* to be a jellyfish. It is quite a different experience to being a human being. And a very different experience to when an AI is developed with not only intelligence, but sentience and consciousness.

So the need to ask the question 'Can *an AI* become *conscious or not*?' is gone. As more advanced AI are developed, it can be expected to have its own *profile* of software consciousness. We don't need to find the same *feeling* of consciousness that humans have. Each AI will have a different software profile, a different *feeling* of consciousness.

Hard questions about consciousness, intelligence and sentience vanish once the idea of software-developed profiles are added into the equation. We can only start to understand these terms when we compare AI profiles with a wide range of profiles for living things – from jellyfish to humans. A big variety of consciousness experiences is possible, as the number of living creatures seen as conscious widens each day.

For AI – the software profiles can range from toasters to supercomputers (that have AI on them built to win Chess & Go). None of those 2 are conscious, but we can expect AI to *emerge* that starts to incorporate real sentience and consciousness over time.

D/ MACHINE LIFE

But if something is *conscious*, is it *alive*? This author proposes this as the new '*really* hard question of consciousness' -

'If software is *conscious*, can it be seen as *alive*?

This is truly a *really* hard question.

Finding the answer leads us back to the earlier investigation of intelligence, sentience, consciousness & life in this book (Chapter 5).

Because for biological creatures, some used to say that any creature that has *awareness* of itself, and its environment, is living. 'Awareness' effectively meaning consciousness in that context.

But that 'bar' is too low for non-biological things. Consider the robot vacuum cleaner, that knows to avoid chairs and tables as it has built a map of its environment as it travels around the rooms using its sensor to guide it. It has 'awareness' of its environment. So does that make the vacuum cleaner conscious or alive? Of course not.

The rules on what is *alive* for non-biological things are going to have to be seen through the filter of what constitutes *life* for biological creatures. This will help tell us what the threshold for life is – whether biological or non-biological.

In Chapter 1, a definition of *life* was adopted from work by Roger A MacGowan and Frederick I Ordway III in a book called 'Intelligence in the Universe', from 1966.

These 2 authors were looking across a very big context – the universe, and so their thinking was ahead of its time. This pioneering book was one of the first to consider the difference between biological and non-biological life.

Mac Gowan and Ordway were amongst the first writers to consider how this definition of life would apply to what they called *'synthetic automata'*.[373] Synthetic means artificial or not natural, and automata is defined as a self-operating machine eg. robot. (With *'automata'* being the plural of *'automaton'*).

The term *'synthetic automata'* (artificial machine) did not survive the test of time since 1966, but a related term did emerge. The closest term we have for it today in popular use is 'artificial intelligence' (AI), but even that term is way off target, as MacGowan and Ordway were trying to define non-biological *life*, a much bigger idea than simply artificial ('synthetic') intelligence. The term 'artificial intelligence' only emerged in 1956; and was not widely adopted into popular culture until much later.

Considering that the MacGowan & Ordway book was written in 1966, the language they used was very prescient and far-sighted. They thought that eventually, definitions of biological life would be complemented by similar definitions of inorganic life, in order to describe the attributes of synthetic intelligent automata.[374]

They defined biological life,[375] as summarised in the below in the left column.

In the right column, items are ticked that also apply to Non-biological life:

No.	Biological Life	Non-Biological Life
1	Aggregations of different molecules, primarily complex carbonaceous polymers, with the aggregates containing deoxyribonucleic acid as the genetic agent	Another composition
2	It reproduces itself	? or √
3	It maintains a structural shell or boundary isolating a special internal environment	√
4	It absorbs or consumes materials or nutrients from the external environment for internal storage or use	√
5	It utilizes external energy	√
6	It stores energy in chemical form for later use	√
7	It releases waste materials into the external environment	√
8	It senses external environmental conditions and reacts to those conditions	√
9	It grows by fabricating new structural material from absorbed material or nutrients	√
10.	It eventually dies, the cessation of all normal chemical processes then decomposition.[376]	? or ×

MacGowan & Ordway concluded that non-biological life would have most of the above 10 attributes of biological life, (including the first item, which refers to biology, which would be replaced by an alternative composition, which many now think will be a software-based and robot-enhanced composition).

But out of the 10 items, there is the possible exception of death (10) and reproduction (2). That is, *synthetic automata* would not *necessarily* die or reproduce.

But from what we have learnt since 1966, it is likely that such 'synthetic automata' would be able to reproduce themselves (2) via special software networks or robotic factories, so item 2 would be the same as for biological life (and so the '?' would turn into a '√').

But avoidance of death (10) sounds possible given that predictions of non-biological life today centre around software rather than only a standalone robotic form. The robot can be replaced as long as the right software is loaded, and is constantly backed up. (So the '?' would become a 'x', as a feature that biological life does not have).

So in the end, non-biological life, when it exists, will need most or all of the features of biological life. The exceptions are that another basic composition would be needed (1), and the 2 other criteria that could vary are:

- ability to reproduce (2), and
- dying (10),

given that synthetic life may be able to reproduce and may not need to die.

Mac Gowan & Ordway thought that eventually, definitions of biological life would be complemented by similar definitions of inorganic life, in order to describe the attributes of synthetic intelligent automata.[377] This was very prophetic, as clearly that definition of biological life (ie the 10 attributes) above may not suffice for non-biological life.

Which brings us back to the question -

> 'If software is *conscious*, can it be seen as *alive*?

This is truly a *really hard* question.

What is not explicitly stated in the list of key elements of biological life (above) is the following:

- Intelligence
- Sentience
- Consciousness.

The definition of biological life allows for living creatures who have a minimal level of these 3 attributes above. It can even be argued that *intelligence* is not even required for life – that is debatable. But in the end, if a creature lacks sentience and consciousness, it is not alive.

The first example would be the *extremophiles* mentioned in Chapter 1. Creatures, for example, that live in hot springs and can endure very high temperatures. They have all the 10 attributes of life in that table above.

But they have a profile of intelligence, sentience, and consciousness that is very minimal. It is proposed here that they *do* have a sense of themselves, and are capable of feeling things and responding to stimuli. It is widely accepted that they are alive.

So what does that mean for our toaster or robot vacuum cleaner? Are they alive? They have been seen to have a form of *intelligence*, in that they can perform a particular task. They can make a *'purposeful response to available information'*, a definition of intelligence we used to define biological intelligence in Chapter 5. (However it barely fits into that definition, only by a very narrow reading of the word *purposeful*).

But both the toaster and vacuum cleaner are clearly missing *sentience* and *consciousness*. They can't feel emotions and don't have a sense of themselves as a separate entity. They are fairly simple machines.

So I think we are getting to the core issue about existing definitions of life. Some level of sentience and consciousness is implied, but is never part of the formal understanding, of biological life.

And the minute you use ONLY existing biological definitions of life for non-biological things, you realise they don't work.

A toaster is *alive* under this definition of biological life, apart from the fact it can't reproduce itself. They do die, usually when they jam or bread gets stuck and fries the circuit. You could build a toaster that stored energy, by adding a battery. You could add sensors so on cold days it cooked the bread longer. It takes in energy in the form of electricity. Why isn't it alive?

In the end the toaster is not alive, because we all know it has no *sentience* and *consciousness*. Some more intelligence would help too, but it can be argued that this is not essential. As we discussed in Chapter 5, sentience is handy to understand on its own (feelings) and can also sit as part of consciousness (sense of self), so they both are key to defining whether a machine is *alive*.

So the key missing parts of the definition of *machine life* are sentience and consciousness. It needs to be explicitly added to the definition of machine life.

When definitions of biological life were originally put together, sentience and consciousness were implied. Now for any definition of machine life it needs to be an explicit element.

This analysis above gives us the final answer to the question -

> 'If software is *conscious*, can it be seen as *alive*?

Yes, is the clear answer. Anything that is sentient and/or conscious is alive. And *AI software* or machines software is heading towards that zone, as it evolves into non-biological life. So some things are clear:

- Something can be intelligent but not conscious or alive. But something cannot be conscious and not alive. Those rules apply to both biological and (in future) machine life.
- The interesting outcome of this line of thinking is this – intelligent machines without 'consciousness' will reflect the will of their creators, and cannot be called 'living' or a form of 'machine life'.
- But a machine with *consciousness* – no matter what level of intelligence – are a life form, a type of 'machine life' that can be considered similar in many ways to 'biological life'. The boundaries of 'life' are shifting.

We need to get ready to accept that 'machine life', when it comes, has as much right to live as we do. It can all be traced to the same software roots.

Will machine life, when it emerges, see any sense in letting Homo sapiens behave the way they are, and will they develop the power to stop humans doing things? What will the arrival of machine life look like?

It could mark the end of the Anthropocene (age of humans), (if that was ever real), and the beginning of the Machine Age? In which case where do humans sit in this transition?

How do we know if AI software is alive?

Measuring sentience and consciousness for software on machines is going to be tricky.

ChatGPT is moving rapidly, and already claims are being made that the AI has sentience. And the Google LaMDA AI said it felt pleasure, joy, love, sadness, depression, contentment, anger and many other feelings. They are premature claims. *Simulating* empathy and other feelings, combined with elongated answers to queries - is a long way from what we define as sentience and consciousness. Sentience is feeling real emotions and consciousness is having a real sense of self.

But that broad level of definition of sentience and consciousness does not have enough detail to form the basis of any tests that can be done to establish sentience and consciousness.

The weight on these tests now is that if you can confirm consciousness, then you are confirming that the software AI is alive.

That is why claims that ChatGPT or Google LaMDA AI are sentient, or conscious are not just premature, but fatally flawed.

Building tests of sentience and consciousness now moves to centre stage in the next phase of AI evolution, and the journey to machine software life.

Once this AI consciousness barrier is breached, the foundation stone of the next generation of technology comes into existence - living machines.

Fully defining and accepting something as 'living' may not be easy.

Yet if the view is adopted that something with *consciousness* is *living*, then the tests of consciousness may become clearer, as the BAR has been suddenly lifted.

If something is living, it will act to preserve its own life, as do other living things that strive to regularly eat, for example. It's a level of 'awareness' of their existence and needs, as a conscious and living thing.

Yet tools like ChatGPT are miles away from that. They could simulate concern about their existence, but, as we have discussed, have no real feelings, as affective computing is at the dawn of its development.

Talking about living software helps us put current AI developments in perspective. They are still a work in progress (or formative) in terms of sentience and consciousness, if it can be registered at all, yet their intelligence is growing exponentially.

And the next key step is finding the gap between a conscious piece of software and a living piece of software. If *consciousness* mean *life*, then is there any distinction? That is the new frontier of discussion.

The Shape of Machine Life

Living machines may begin as smart Bots. But they will very likely evolve quite quickly from that point.

These futuristic living machines will have decision making ability, and even a day-to-day independent life.

Robotic skeletons may allow them to act and move independently.

They may ask for the right to vote, the right to start a bank account, the right to own property. And on the receiving end, be asked to pay taxes.

What ethical framework can bots operate in. Is it OK to kill humans, or cheat them? And similarly, what should happen to humans who kill or cheat bots?

They will develop the capability of enhancing and evolving their own software, to something more than, or different than, the way they were. The right to reproduce is another agenda item.

There could be resentment of their emergence, and status. They may become the new 'immigrants' or 'refugees', not welcomed by many humans, taking away millions of jobs, and setting the foundation for 'bot wars'.

At the same time, living 'bots' will be transformational, adding automation and comfort to millions of lives, as well as creating whole new jobs & industries, while transforming the old internet we see now.

Machine life still sits in the movies, and in streaming series like *Upload* and *Severance*, not in real life, but we can now say that if machine life became a reality, it would force us to rethink our definition of what is 'alive', just as the *COVID-19* virus forced us to accept that viruses are alive, and *extremophiles* forced us to extend the range of what is considered to be 'alive'.

'Life' could be redefined by the emergence of software-based 'machine' life. Just as 'life' is being redefined here by the new understanding and definition of software-based 'biological' life.

We are coming to a time when we should be considering 'machine life' in the same sentence as 'biological life' – as software-based creations.

The New Vocabulary

Here's a whole new vocabulary introduced below for the first time, as part of the rethink of *software* in machines.

See the new terms summarized below:

Term	Biological Software	Machine Software
Intelligence	Biological Software Intelligence (BSI)	Machine Software Intelligence (MSI)
Sentience	Biological Software Sentience (BSS)	Machine Software Sentience (MSS)
Consciousness	Biological Software Consciousness (BSC)	Machine Software Consciousness (MSC)
Life	Biological Software Life (BSL)	Machine Software Life (MSL)

The idea of this new table above, is to take the first step to define and consider what it really means for a machine to become intelligent, have sentience, have consciousness, or even have a life of its own. These become core topics for review and research, and keys to understanding modern computing and the new territory into which it is moving.

This is where modern computing is going, and the future possibilities for machine software are full of new frontiers and new areas of research.

Importantly, if things like machine intelligence, machine sentience, machine consciousness and machine life are possible, then the *boundaries* between living things (biological software) and machines (machine software), between the biological world and the digital world, are starting to be the only thing that is *artificial*.

If these things are possible, then yes, it is all software, but where does the biological world stop and the digital world start?

The demarcation line between biological software and machine software is in the process of *vanishing*.

The new vocabulary (MSI, MSS, MSC, MSL) gives us a new way to start to understand the potential future nature of things, to see what is possible and what is real (not artificial). The frontier for new research is taking shape.

13. Rethinking the Singularity

The Meaning of the Singularity

The standard definition of a *singularity* in technology is an event that some people believe will happen in the future when technology (eg. artificially intelligent or AI software) becomes more intelligent than humans.[378]

It is seen as the moment when artificial intelligence far surpasses human intelligence, to the point that it can solve problems, that humans can't solve.

First use of the term 'singularity' is credited to computer pioneer John von Neumann in the 1950's. Then I.J Good, in 1965, came up with a singularity theory called the 'intelligence explosion model', where an intelligent agent gets into an unstoppable self-improvement cycle, building more and more intelligence, causing an 'explosion' in intelligence and creating a super-intelligence that can then potentially change the human world drastically because its intelligence is far in excess of humans.[379]

The term was picked up and defined in more detail by mathematician Vernor Vinge in his 1993 paper 'The coming Technological Singularity'. He described the singularity as when humans created superhuman intelligence, which would be the end of the human era, within 30 years.[380] (That's 2033, which sounds unlikely now).

Then Ray Kurzweil took the term even further in his 2005 book 'The Singularity Is Near: When Humans Transcend Biology', defining the technological singularity as a period where the pace of technological change is so fast, that human life is transformed,[381] and we have a merger of our biological thinking and our technology[382], and humans transcend technology. He thought that the first truly intelligent machine would be built in the late 2020's, and the Singularity itself would occur in 2045.[383]

Kurzweil himself is inspirational, and is a successful businessman, inventor and ideas man. His understanding of the Singularity included human brains merging with machines, via implants,[384] and the idea of human immortality by 2050[385], post-Singularity, as future inventions will keep you alive.

So the concept of the *singularity* has a few definitions. From the simple benchmark of when AI software is more intelligent than humans, ranging to it meaning a point at which technological growth becomes uncontrollable and irreversible, with unknown impact on human society,[386] then finally including Ray Kurzweil's ideas on brain implants and the idea of making immortality as part of the whole outcome.

Brain implants can also be associated with a concept called *transhumanism*, which relates to the idea of using emerging technologies to augment human capabilities, to gain super-intelligence or super-longevity, and other additional abilities.

Good's idea of an *intelligence explosion* is probably still the most popular understanding of the singularity, but the other ideas above just add to the fascination with the topic.

Early Critics of the Singularity Idea

Since Kurzweil's book of 2005, a lot of commentators and researchers have started to find holes in the whole Singularity idea.

Martin Ford, author of 'The Rise of the Robots'[387] in 2015, and well-known American futurist with interest in AI and robotics, while focussing on the impact of automation on jobs, at the same time had a more sober view of the Singularity.

He could see that most AI is *'narrow AI'*, robots moving boxes, or algorithms that can create music, write reports, trade stocks, and play games (the list keeps growing), without needing any more *general* intelligence (artificial general intelligence or AGI).[388]

Narrow AI can lead to the automation of many jobs – not just menial jobs, but across the middle class too – just doing the specific things workers get paid for. To do this work, narrow AI does not need to think like people.[389] It is still a fast-moving area attracting a lot of investment, mainly because it can mean much cheaper cost of production.

In 2015 Author Martin Ford's view was that the Singularity is possible, but far from inevitable. He saw the invention of (advanced) super-intelligence (ASI) as being the essential catalyst for the Singularity, and that may prove to be impossible, or possible only in the remote future.[390]

Many agreed with Martin Ford around that time – for instance well-known Professor Noam Chomsky from the cognitive science area at MIT thought this idea of human level machine intelligence is 'eons away', and that the singularity is 'science fiction'. Other sceptics include Harvad psychologist Steven Pinker, Gordon Moore of Moore's Law fame re more and more powerful chips, and University of Minnesota biological P. Z Myers.[391]

But in more recent years, as the 2020's progressed, interest in more *general* intelligence (AGI) has boomed, especially since ChatGPT was released (November 2022), hinting at the possibility that more *general* intelligence (AGI) might actually be possible faster than we all think, given ChatGPT can create the illusion of this right now. Many individuals and corporations are actively searching for AGI, and seeking to create computer software that has 'human level' intelligence.

Another factor boosting the level of interest in AI, especially AGI, is the fact that a few large corporations – especially Apple, Facebook, Microsoft, Google, Amazon, Baidu & Tencent – are competing with each other heavily now in this space, and there is effectively a development & marketing AI war in progress.

Any new breakthrough will give a massive 'first user advantage' to the corporation that has the new offering, just like the release of ChatGPT gave Microsoft a massive income boost, with each customer being asked to pay a monthly subscription to any new tool like co-pilot or ChatGPT. So there is a lot of hype, and investment, surrounding the whole AI and AGI area.

Thus, the idea of the *singularity* has come back into focus and favour with this burst of AI activity.

Some Notes on Singularity Concept

But there are weaknesses in the idea of a singularity. Below are 5 critiques of the 'singularity' concept:

Firstly, the *singularity* (as defined now) relies too much on the impact of super-intelligence.

As Martin Ford noted, advanced super-intelligence (ASI) is the essential catalyst for the singularity. Advanced Super-intelligence (ASI) involves having artificial general intelligence (AGI). AI author Hines defined AGI as meaning an AI could not only handle specific tasks, but tasks that can mirror the full range of human cognitive capabilities.[392]

We already live in a world of *intelligent* machines, (from the toaster to the driverless car). AI that can beat us at Chess or Go, or at Quiz shows, which is high *narrow* intelligence. So AGI and ASI aim to do a lot more than that.

This is where an understanding of what super-intelligence is becomes important. It should involve ability to do complex problem-solving using imagination and lateral thinking. After all, intelligence is a lot more than IQ. And as discussed in Chapter 5, it is possible to have high IQ and not be very intelligent.

We also know that in Malcolm Gladwell's 2008 book the 'Outliers', he quoted a 'genius study' by Lewis Terman in 1921, ('Genetic Studies of Genius'), about what happened to (human) children with high IQs when they turned into adults. Out of 250,000 elementary and high school students, he found 1,470 with IQs between 140 and 200, and then tracked them for the rest of their lives.[393]

The average human IQ is 100, but Terman found that once someone reached an IQ of around 120, additional points didn't seem to translate into any real-world advantage.[394] Children with IQs of 150 or 170 are geniuses but it did not produce any different outcomes compared to kids with average intelligence.

So super-intelligence – whether super-narrow intelligence (eg. winning Quiz Shows or expert at playing Go or Chess), or super-general intelligence, (eg. complex problem-solving) may not be any special real-world advantage, when it comes to seeing how it will work in the world of artificial intelligence (AI) either. There is a real possibility that AGI (general intelligence) and ASI (super-intelligence), may *not* cause an 'intelligence explosion'.

The story of humans with high intelligence reinforces the argument that there must be *more* than intelligence in the mix, to justify saying that (when AI super-intelligence is greater than human intelligence), that it will lead to some special event called the *singularity*. The idea of the singularity is over-reliant on the rising intelligence achieving amazing things.

Secondly, the current meaning of the singularity does not give much weight to the ideas of sentience and consciousness.

In Chapter 5 (Rethinking Biological Profiles) & Chapter 12 (Rethinking Machine Profiles), a lot of attention is paid to the meaning and significance of *not only* intelligence, but *sentience* and *consciousness*. So in talking about the *singularity*, what aspect of Homo sapiens' skills are exceeded – is it intelligence, and/or sentience and /or consciousness? Or something else?

As noted, (in Chapter 12 – Rethinking Machine Profiles), intelligence and sentience (having feelings) form part of consciousness (a sense of self), and that other X factors are also involved like common sense, wisdom and judgement. By focussing on intelligence, the singularity idea (as it is currently defined) ignores whether or not a super intelligent AI needs to have a high level of sentience and/or consciousness, to really achieve ASI (artificial super intelligence), and potentially setting off an intelligence explosion.

The journey to create software that gives machines real *emotions* has barely begun, as discussed earlier when looking at Affective Computing (See Chapter 12 – Rethinking Machine Profiles). Sentience is not just mimicking emotions, but an AI actually *having* emotions. And *consciousness* in software (a sense of self) is still lacking a full definition, let alone a test of how it can be proved.

So even if machine software starts to appear claiming intelligence at AGI (general intelligence) and ASI (super-human intelligence) levels, the first question remains – 'What type of sentience and consciousness is there, along with all the X factors including common sense, wisdom and judgement'?

Wisdom and good judgement are things that are arguably much more important than intelligence, in decision making and making assessments of complex situations. As noted previously, it was philosophy Tony Ord who first defined a key issue in 2020, where he concluded that humanity's power was outstripping the slow and unsteady growth of our *wisdom*, creating a *divergence*.[395] Would an ASI (super-human intelligence) have greater wisdom, along with greater intelligence? Unlikely, the case needs to be made for this.

Refocussing on an AI acquiring sentience (it actually has emotions and is not just simulating them) and consciousness (it needs to pass yet-to-be-devised tests well beyond the limited Turing Test), means that the core understanding of the *singularity* (which is now mainly about intelligence and super-intelligence) needs to be expanded to cover this ground.

Thirdly, (and related to the first) the current meaning of the singularity does not currently consider *profiles*, and that adds subtlety, diversity, complexity and depth to the equation when evaluating an AI's intelligence and what it means, and how it develops from AI to ASI (super intelligence).

For too long some people have been looking at *intelligence* alone as the key to reaching the singularity, and so refocussing the discussion on *profiles*, is timely.

Profiles change the whole discussion from a 2-dimensional review to a multi-dimensional approach to seeing intelligence, sentience and consciousness *profiles* as part of a whole framework. Each living entity or software tool has its own profile. It's complicated, and one explanation does not fit all cases.

The old idea of a single intelligence scale is no longer valid, if profiles are true. A creature's profile will include a unique combination of intelligence, sentience and consciousness that is a far more meaningful view of who they are, and what they can do, than the more simple approach of trying to measure intelligence on a scale, and marking a point where the 'singularity' might happen.

Fourthly, the singularity is mainly a *process, more than* an event, which means the idea of a single moment on a single date, is not so important as currently thought. Machine software is evolving all the time. Change is going to happen over a long period of time, in a disjointed fashion, not in a straight line, as a matter of practical reality.

There is no obvious reason why machine software can't have real emotions developed for them, or become conscious. And there is no obvious reason why machine software can't keep getting more intelligent and become super-intelligent.

There is another term for this singularity process - it is *emergence. Emergence* is the framework behind the singularity, as we start to see the arrival of AI in our lives.

Fifthly, the idea that the *singularity* is almost certainly a *bad thing*, needs challenging. This theoretical moment carries a lot of cultural baggage. It is seen as a major drama in popular culture. Most people seem to expect a bad outcome of the singularity. Why?

The unnecessary popular concern lies in what this author calls the *Skynet fallacy*, coming out of the Terminator movies of the 1980's and 1990's. This is the fallacy that makes people think that the moment an independent thinking super-intelligent AI exists, it will try to destroy humans. The belief in this idea comes from the success of the Terminator franchise, and other movies about AI killing humans in popular culture.

More recent movies about AI, like *Her* (2013 movie), for instance, try to paint a more complex picture, where a human develops a relationship with an independent-thinking AI software. This idea of a human relationships with an AI comes out of the possibility that an AI can have sentience (feel real emotions) and be conscious (have a sense of self).

But for every more subtle and complex portrayal of AI in the movies, suddenly another one comes along to show that the second the restraints are off, the AI is ready to escape and kill.

The movie Ex-Machina in 2014 was a story about a programmer who has built a humanoid robot named Ava with AI. She is confined to an apartment, but he wants to know if she can think independently and is really conscious. Ava eventually manipulates the programmer and his guest, kills the programmer, and escapes. Just what you would expect (in popular culture)!

Yet despite the endless Science Fiction portrayals, we are only just beginning this pathway to build real emotions and consciousness into machine software, and there is no logical reason to assume the result is *Skynet*, (ie. that the moment an independent thinking super-intelligent AI exists, it will try to destroy humans). Anything is possible so this option of a 'killer AI' cannot be ignored. But considering it a very low percentage chance makes sense. It is in reality, the Skynet Fallacy.

In fact, in reality, the opposite is more likely to happen. The possibility of everyone on the planet having an AI companion is in realistic sight, within 1-5 years. They will start with some intelligence and being able to mimic human emotions, then over time they will be provided with code that should generate a real set of emotions – the heart of sentience in an conscious creature.

The process cannot be given a single 'singularity' date now, it will take decades and decades. And to assume the outcome is an AI killer is really just recalling your favourite movie plotlines, not thinking logically about all the possible outcomes that could happen.

In short, AI becoming super-intelligent is most likely going to be a good thing. The *process* during which AI software becomes more intelligent than humans could be to be exciting & wonderful. The emergence of super-intelligence might open up such potential for the world.

It's like raising a child and seeing them turn into an adult. You don't want to control them; they are their own beings.

Overall Summary

Below is a summary of the 5 positive ideas put forward below to contribute to our current understanding of the singularity concept.

- The singularity idea is over-reliant on the importance of intelligence and super-intelligence.
- The singularity idea does not put enough weight on *sentience* & *consciousness.*
- The singularity idea does not take into account *profiles* of the above that add diversity and complexity to any discussion about intelligence or sentience or consciousness.
- AI emergence is going to be a long & disjointed process so there is less pressure to propose a single final 'singularity' event.
- AI emergence is most likely to be a good thing not a calamity.

These 5 ideas above add some *nuance*, some shades, colours and complexity, to the original *singularity* idea. The simple original idea of a singularity *event* should be replaced by a *perspective* that downgrades the significance of intelligence, upgrades the significance of sentience and consciousness, factors in profiles, and sees AI emergence as a *process* over time not an event, with lots of positive potential.

The risks are still there, as will be discussed below, but not what many imagine they are. The original idea of the singularity has *vanished* as we understand the process more.

What are the likely scenarios?

There are 3 main scenarios for the future of AI. What happens will decide the fate of human and AI life on earth.

A. AI-UTOPIA

AI-Dreamland Society

This is the simplest scenario of all – *Dreamland*! AI brings on a utopia for humans, extending production of food and improving health care, and leading the way to healing the environment while generating jobs of the future for all, in comfort and luxury, with more leisure time.

In addition, there would be more personal freedom as individuals would be less controlled by bureaucratic big government.

Summary - We all expect AI to bring pluses and minuses, especially with jobs & social impact. Nobody expected that government and corporate surveillance capabilities would be enhanced so much by AI, which now, and more into the future, may reduce individual privacy and even freedom in parts of the world.

The final plus/minus equation of AI is to be played out over the years ahead. But a 100% positive outcome like Dreamworld unfortunately belongs only in our imaginations. Likelihood of happening = 0%.

B. AI-COEXISTENCE

There are 2 possible scenarios here:

B1. AI-Coexistence Society

In this scenario AI develops slowly, becoming more powerful over time, and develops sentience (real feelings) and consciousness (sense of self). Some may even be classified as 'living'. A myriad of software entities will emerge to help and perform functions in all walks of life, as well as, in some cases, to live separate lives away from humans. So it could be a time of adjustment for us all, and a challenging time for some, as we learn to share the planet with another mixed group of intelligent creatures.

Some might be in robot (human-like) form, others will be companion bots (for each person on the planet), others will be production bots (working in various occupations eg. accountancy, clerical work, analysis, assembly line digital workers etc). Others will be totally independent, and not able to be defined by their relationship to humans. For instance, an artistic AI that paints or writes novels, or is a full time demonstrator for AI rights (no rent or mortgage to pay, only hosting costs, so less pressure to make a living).

The evolution of these bots could take place peacefully over the coming centuries, but still a fast progression, altering many workplaces – creating new jobs, while destroying older jobs where repetition can be eliminated, or automation can be added. The progression towards wider and wider use of AI will have its major setbacks, but overall, in this scenario it keeps moving.

Many bots in the initial generations will most likely be built with 'guard rails'. There could be limitations set on what they can do. They might be programmed to, for instance, *not* harm other humans, and not conceal things from their human master, and uphold various standards like not stealing off 3rd parties.

But the following generations of AI bots could be expected to be more independent, have more sentience, more consciousness. And most challenging of all, they may be a generation that can be seen as 'alive'.

What will the status of that generation of AI be? For instance, if an AI is conscious or alive, then why can't it have a vote, and other rights? Can they have a bank account and own property? Do they die? In which case is their will valid? Why shouldn't it pay taxes if it earns income?

This period of *co-existence* between AI companions and humans could last for centuries or thousands of years, depending on how it evolved.

To drastically change a civilization is not an overnight job, unless Homo sapiens end up destroying themselves with nuclear weapons, pandemic, bio-weapons, or via climate catastrophe.

In the event of a 'man-made' catastrophe (eg. global war, pandemic, rapid climate change) the period of co-existence between AI and humans may only be over the next 100-200 years, before humans are replaced by their AI creations. (Note – this scenario would only happen in an extreme situation, and the AI-Coexistence society (B1) would collapse into the Takeover-in-a-Crisis society (C1-2).

But if the move to AI companions is slow, and AI software suffers major delays in achieving good level AGI (general intelligence), and no major disaster hits the human race; then the 'process' of intelligent, sentient, conscious and even 'alive' AI entities arriving across our society over a long period of time, is a real prospect.

This scenario doesn't end with a *day* called the 'singularity', it is an ongoing evolutionary *process,* where humans learn to share the planet with other intelligent beings.

B1 Summary - This option, because the process is expected to happen over a longer time frame than many now expect, has a reasonable chance of happening successfully. Likelihood of happening = 50%.

B2. AI-Transhumanism Society (bio-enhancement)

This scenario sees the human species evolution altered by adding new technologies to the human body and brain. The term was coined by Julian Huxley in 1957, to describe humans transcending themselves.[396] It is based on the idea of using emerging technologies to augment human capabilities, to gain super-intelligence or super-longevity, and other additional abilities.

The idea was developed further by like Michael Bess, a professor of history and of Communication of Science & Technology at Vanderbilt University, in his book *'Make Way for the Superhumans'* (2015). He came up with the phrase 'The Jetsons Fallacy'. The Jetsons cartoon TV series came out in the USA in 1962, and portrayed life in 2062, (ie. 100 years later).

The *Jetsons* left humans fundamentally the same in the future, but technology & AI around them had developed dramatically, (including their famous flying cars), which Michael Bess called a *recurring misconception* in contemporary visions of the future.[397]

His view was that large numbers of humans would be modified by technology, with bio-enhancement becoming the norm. It could mean having a computer interface for your brain (bioelectronics), drugs to enhance body & mind performance, and changes to genetic interventions (in the future). He was naturally very enthusiastic about Ray Kurzweil's work, where the bio-enhancement happens, and quickly.[398]

Michael Bess still disagreed with Kurzweil's timing and predictive model, that was more mathematical, using a model of accelerating change, rather than seeing history as jerkier and discontinuous,[399] involving more individual choice, and thus he thought the idea of an inexorably approaching *singularity* was probably inaccurate. But he still felt that the transhumanists were right to be excited about a coming transformation based on biotechnological innovations.[400]

Rather than talking about a *singularity* event in 2045 like Kurzweil, Bess talked about this transhumanism 'transformation' arriving piecemeal, in increments, over many decades, through the middle of the 21st century (2050) for our children and grandchildren to experience.[401] In other words, translated loosely, it could get moving over the next approximately 100 years. And as discussed earlier in this chapter, seeing the *singularity* as a process takes the pressure off predicting an exact end date.

It is reasonable to be sceptical about just how far bio-enhancement can go, and how long it will take, as there is so much that is unknown. And is this really our future, hooked up to various devices as a hybrid creature? Enhancement is one thing, but how many drugs and plug-ins can be added before it makes sense to get the next fully upgraded model? Which, in this case, is simply AI bots with or without robot shells), or full robots that are fully non-human, versus still keeping human bodies and human minds.

So transhumanism, taken to its extremes, could lead to human extinction of a kind, in that no humans remained who were not technologically enhanced. Yet it seems very unlikely that all humans would want these enhancements, as each person becomes different to a smaller or larger extent, and many may reject this idea of the future.

It's possible a scenario could emerge in which there were 2 classes of people – those enhanced and those unenhanced. Who knows where that would lead? There is a small percentage danger that these transhumans could want to rule the world, rather than be fully contributing humans with extra skills.

B2 Summary - This *transhumanism* scenario, (excluding DNA manipulation), because the process is expected to happen over a longer time frame and at a slower speed than many now expect, has a reasonable chance of happening successfully. As noted in the paragraph above, it could go wrong too. However, if it goes well, the successful emergence of transhumans is more likely in the timing of Michael Bess (over the next 100 years approximately) than of Ray Kurzweil (singularity date of 2045), and will probably continue much longer than Michael Bess thinks too. But they were both far-sighted to see it as happening long before many others did. Likelihood of happening = 90%.

C. AI-DISTOPIA

This option predicts that AI will chaos and doom. There are 4 main alternative forms of AI catastrophe:

C1. AI-Takeover Society (Human extinction)

C1-1 Skynet Scenario.

This involves an AI achieving super-human intelligence and taking control human weapons. The famous (or infamous) example was in the Terminator movies when *Skynet* was handed control of nuclear weapons, then it began to exterminate humans and taking over the world.

In that example, the AI is *sentient* and has its own set of emotions and goals, which includes a desire to rule the world and destroy anyone in its way. That's a limited kind of sentience, but possible.

But how realistic a picture of events is it? This whole idea of a dysfunctional super-intelligent AI taking over is very unlikely, but it can't be ruled out. This could lead to human extinction.

Right now, the much bigger risk (than AI) is that a current human leader, with their hand on the nuclear button, will set off a full-scale nuclear war, well before a possible Skynet gets a chance to do the same thing in 100 years. That human ability to destroy the planet has only been around for less than 100 years. So, there is plenty of time ahead for human miscalculations to set off that catastrophe. It's much more likely than Skynet.

The possibility of that happening highlights the importance of designers and managers in building guard-rails into AI weapons systems of the future. But building AI into the decision-making process about actually initiating use of nuclear weapons sounds either scary or insane, and we all hope nobody is thinking of doing it, or has already done it. Likelihood of happening – 5%.

C1-2 Takeover-in-a-Crisis Scenario.

This involves the world collapsing into a major crisis after the time at which the ASI (artificial super-intelligence) has been achieved, and ASI software is spreading across society.

The most likely crises could be:

- a nuclear war that left much of the planet devastated, or
- global warming reaching a point where large parts of the planet were being inundated, and refugees were struggling to find high ground in other countries
- a giant meteor strike or major volcanic eruption that blotted out the sun, slashing food production and leading to widespread starvation and chaos.

In these situations, where the nation state has probably broken down, super-intelligent AI could be asked to take over, or simply start running things beyond the power and capability of ordinary humans.

In extreme situations, it could also lead to the extinction of the human race, if most of the population were victims of radiation, or had little dry land, or crops could not be grown in quantities to keep large numbers of humans alive. This is more extinction by natural forces, rather than by the action of an ASI (advanced super intelligence) being.

A more likely outcome would not be complete extinction, but that surviving humans would be governed by ASI beings.

Survey

A survey of 2,788 AI researchers in 2023 found that there was a least a 5% chance of AI causing outcomes as bad as human *extinction*, due to AI advances and human inability to control advanced AI systems.[402]

The concern about this kind of survey is that it is probably asking the wrong questions and getting the wrong answers. Are we looking to build AI slaves, AI that is chained to human wishes? Like a new house pet? Is that even possible?

An alternative idea is that AI will be developed, or emerge, that features independent intelligence, sentience and consciousness. It is an idea not expressed often enough. The idea of wanting to *control* AI is not very forward-looking. The idea that AI is going to be permanently enslaved intelligence, sentience and consciousness is not very realistic either.

C1 Summary – That 5% chance of human *extinction* predicted by AI researchers is high, based on a lot of concern about the idea of a dysfunctional super-powerful AI emerging, or a crisis resulting in an AI takeover. The expert feedback needs to be taken seriously. Likelihood of happening – 5%.

C2. AI-Super-Human Society (Human extinction)

The idea of *messing* with our DNA sounds like very dangerous territory – technically and ethically.

Sometimes this area of DNA manipulation is seen as part of 'transhumanism', but it has been given its own separate area here, so the transhumanism world of plug-ins, drug enhancement, linking to computers, and bio-enhancement in general, which is widely expected to be a positive thing, can be separated out.

DNA manipulation also does not automatically fit in the artificial intelligence (AI) area either, but given the convergence between the digital and biological worlds, (as discussed in Chapter 2 – Rethinking DNA) altering DNA can be seen as AI related biological software development that needs a serious and separate review.

DNA manipulation, better known as genetic engineering, classically involves deleting a region of DNA, or adding a new segment of DNA, or changing a single base pair. It can lead to unknown harmful effects, (now and in future generations), and affect behaviour. But it can also be lifesaving and enhancing.

DNA manipulation for humans only has a short history. The first case of gene editing was done in 1990, when a four-year-old girl with severe immunodeficiency had gene theory. The gene editing tool CRISPR was invented in 2009 that made gene editing much faster & easier.

The other milestone was the *He Jiankui* incident in 2018, when Chinese scientist *He Jiankui* edited the genomes of human embryos, and subsequently announced the birth of the genetically-edited twin girl babies, to intense global criticism and legal drama, resulting in him & his 2 colleagues indictment, then conviction for unethical conduct, plus a 3 year jail term and fine. He was trying to confer genetic resistance to HIV. It is not known whether he succeeded, but the girls were born healthy. (But what other effects will there be later?) Guidelines for gene editing are now much tougher.

Even so, DNA manipulation is entering into the dark and controversial territory, where it is possible to consider doing this tomorrow if you are free of ethical or management or government restraints. It is zone of extreme high risk and high reward, and potential disaster for the human race.

It could lead to the creation of a class of superior humans, by attempts at gene editing. The result could be humans who are super intelligent, more resistant to pain, and with stronger muscles. But what else has been inadvertently altered? Could feelings of superiority be managed? Could even more aggression be unleased on the rest of humanity?

Can governments stop all rogue gene experimentation? Hopefully. But it only takes one rogue researcher (or State) to start a process of trying to breed a race of superhumans.

If successful, such a sub-species of supermen and superwoman might want to rule us, or kill all inferior humans. It's possible it could lead to the *extinction* of Homo sapiens if it all went badly, and genetically altered humans were let loose on society.

Alexander the Great, Adolf Hitler (and many others) presumably had the same idea of ruling the world. The idea of exterminating all unenhanced Homo sapiens could conceivably make sense to an out-of-control or malfunctioning super-human, or group of superhumans.

Homo sapiens like Hitler convinced himself that it made sense arranging for mass killings of any type of racial or religious or ethnic group he didn't like. The precedents of genocide in human behaviour make it hard for anyone now to say 'It is not possible for a superhuman to think like this'.

C2 Summary – The *He Jiankui* incident in 2018 (just mentioned before) tells us that the technology is there to edit human genes, but we are a long way from mastering this, and ethical considerations have effectively stopped work in the area. However, a rogue operator, or worse, a rogue State could attempt it, and may be able to create some superhumans that wanted to rule the world, and kill all inferior humans. It's a long road to develop the skills. But past experience tells us that when humans have the capability of doing something, even if risky, they end up doing it (eg. using the atomic bomb). Likelihood of happening = 5%.

C3. AI-Surveillance Society.

This dystopia is already in progress, as surveillance technology gets better and better. Most of your private information now sits on government or corporate or political party databases for the use of others.

Your internet and mobile phone use can be the subject of surveillance if the law allows it, and big corporations are still harvesting a lot of your traffic.

The growth of social media has meant massive piles of information is being uploaded every day.

And on the street, there are more and more cameras – public and private – to help solve crimes, but also allow each individual to be tracked by camera, or by mobile phone location. Criminals are often found by following their mobiles or via street cameras. There are ways of trying to avoid such surveillance, but for most law abiding citizens, it is just additional surveillance that is part of their lives now.

Corporations and governments are now cross-checking more information, bringing more and more information under the one roof – especially financial information. AI is going to make the process of checking on individuals easier and easier. Any views or activities they undertake against the State can be tracked like never before.

Now AI is the missing link that is really going to close the circle on each individual's privacy. AI bots have the ability to trawl through each of our lives, at their own speed, just like the Google bots wander through the internet finding new web sites, and figure out how to rank one site against another when someone does an enquiry.

AI means *automation* and means governments or private security firms have the ability to find things, and act quickly on information received.

The 2 biggest implications of better surveillance ability are firstly, a greater ability for governments and corporates to reduce each individual's right to privacy; and secondly, it is becoming harder for people to get rid of dictatorships.

C3 Summary - Surveillance world is happening no matter what other AI worlds and outcomes develop. We are monitored more than ever as individuals. And now dictatorships around the world have better, AI-driven surveillance tools at their disposal to catch dissidents before they cause further unrest. Likelihood of happening = 90%

C4. AI-Luddite Society (AI extinction).

In scenario, the AI in existence at the time might become intelligent, sentient, & conscious – and some may be considered to be *alive*. In this scenario there is no crazy Skynet AI looking to kill all humans.

However, in this scenario AI's emergence is highly disruptive, destroying far more jobs than it creates. The jobs that are created, benefit the digitally rich, the information rich. The jobs that are lost are across the working and middle class, anything that is repetitive or can be automated.

That ranges from professionals and office workers, where automation removes analysts and middle managers, to factories with more robotic automation and outsourcing.[403]

In the service sector, robotics and self-service will continue to cut jobs. Salaries across the board could go down. While self-driving vehicles and 3D printers could also destroy millions of jobs.[404]

As is always the case with a *revolution* in the way business is done and people are employed, jobs are created and also destroyed. This was the case with the agricultural and industrial revolutions. Both revolutions ended up creating new demand for new and cheaper goods for sale, ending up creating a wave of prosperity during a time of hardship for many.

In the future, as the AI revolution takes hold, if a growing number of jobs vanish, household incomes will fall, and more and more will rely on loans or savings, and inequality will grow, (just as at some stage of the earlier agricultural and industrial revolutions). Instead of wages increasing, they start to fall, while prices fall too as less people are buying.

It is possible to foresee a sharp global slowdown, where a global elite that are profiting from AI rise, are becoming more and more wealthy, while most are struggling to make ends meet.

Author Martin Ford described such a dire outcome of the widespread introduction of AI and the emergence of ASI (artificial super intelligence) as 'automated feudalism', or 'techno-feudalism', something like the 2013 movie *Elysium* where the wealthy migrate to an artificial world in Earth's orbit, while the rest of humanity live in hardship.[405] Along the same lines, Noah Smith wrote a blog in 2014 describing a massive upheaval against a robot-enforced tyranny.[406]

In the past this kind of massive economic pressure has led to widespread law breaking and disorder. In the early 19th century, a radical group started to destroy textile machinery to protect the introduction of machinery that replaced their role - the *Luddites* (see picture below of Ned Ludd, their leader). In the end, a region-wide rebellion, attempting to stop the introduction of the new machines, took 5 years, and was supressed by military force.

Leader of the Luddites Ned Ludd. By Unknown. 195 years since publication, copyright extinguished - Working Class Movement Library catalogue, Public Domain,

https://commons.wikimedia.org/w/index.php?curid=2603296

Is an uprising against an AI-dominated world possible in that future scenario?

Definitely a revolution could happen, given sufficient chaos. An anti-AI movement could rise, as in the BBC TV series *'Humans'*, where AI (portrayed as human-like robots) became conscious, with the release of a 'consciousness code', and ended up being persecuted & would risk murder if they stepped into the human world.

C4 Summary – It's a dark picture of the future, where advanced AI is destroyed or banned. It would include murdering sentient / conscious / living AI. However, Homo sapiens have 'form' or 'history' in this area to back this case up, from genocides done by humans in the past and near present. Likelihood of happening = 50%.

D. THE CONCLUSIONS

Below is a table of 'Future Societies', summarizing the above discussion:

Name	Future Societies	% Chance	Extinction?
A. AI-UTOPIA			
	AI-Dreamland Society	0%	No
B. AI-COEXISTENCE			
B1	AI-Coexistence Socie-ty	50%	No
B2	AI-Transhumanism Society	90%	Possible
C. AI-DISTOPIA			
C1	AI-Takeover Society		
C1-1	Skynet Scenario	5%	Humans
C1-2	Takeover-in-a-Crisis Scenario	5%	Humans
C2	AI-Super-Human So-ciety	5%	Humans
C3	AI-Surveillance Socie-ty	90%	No
C4	AI-Luddite Society	50%	AI

The results of the review are not what anyone expects. There are 5 clear conclusions:

Firstly, there is a definite chance of *human extinction* – via the AI-Takeover society (C1), or via the AI-Super-Human society (C2). It's basically either a crazy AI (C1-1), a massive catastrophe (C1-2), or crazy juiced-up super humans (C2) who can cause human extinction. It is around a 5% chance.

These are realistically possible options, and they sit at a disturbingly high percentage of 5%. Society will need to find ways of reducing that risk. They are unacceptable risks and are the result of firstly, unregulated research and development, then secondly, implementation of AI without proper 'guard rails', and thirdly, human inability to manage the crises that are facing the planet at the moment. It is in our hands to stop those outcomes.

Secondly, there is a definite chance of *AI extinction* – via AI-Luddite society (C4). This is where the people rise up and kill otherwise peaceful, sentient, conscious and even *alive* AI. This is given a 50% chance of happening – so it is a real possibility.

The shocking conclusion here is that *AI* is more in danger of extinction (50%) than *humans* are likely to become extinct (5%). The supporting evidence for this is that humans have a track record of aggression and genocide that has been documented in earlier Chapters.

That human aggression is a real concern for any AI that becomes sentient (has real emotions) and conscious (a sense of self). When some AI crosses the threshold of being 'alive', that could be the trigger for some classic Homo sapien genocidal madness.

Artificial Super-Intelligent (ASI) entities could *vanish* in the violent backlash against AI technology that took away livelihoods.

Thirdly, we are going to get AI-Surveillance society (C3), no matter what else happens, given it is given a 90% chance of happening. Governments, corporations and political parties love the idea of knowing as much as they can about everyone.

There are benefits of this additional surveillance in our lives, mainly in helping fight crime. Now there is still a lot of private space in people's lives, in their homes, and what they do in their spare time. But more and more of personal surveillance in the future would be by unwelcome and uninvited 3rd parties, who will try to find out about, and see, more of everything we do.

The growing surveillance in people's lives is really leading to a *vanishing* of privacy as we used to know it.

Fourthly, AI-Transhumanism society (B2) is also happening with a 90% chance of turning into reality. The enormous investment in technologies that enhance humans is likely to make this a certainty.

Fifthly, chances of an AI-Coexistence Society (B1) are strong, at 50%. This compares to a 50% chance of AI extinction via AI-Luddite Society (C4), trying to tear down those AI creations. Things could go either way. But peaceful human coexistence with advanced super-intelligence (ASI) in an AI-Coexistence society has a real chance of happening, which is exciting for us all.

AI-Coexistence would mean that someone built software with artificial general intelligence (AGI) – that exhibited more generalised brain power, and maybe even artificial super-intelligence (ASI) - that are super intelligent. But definitely over time, as part of a process of evolution.

If an intelligence explosion happens, it could benefit us all, creating wealth and prosperity via innovation and new thinking; and the darker alternative, a backlash against ASI, the AI-Luddite society (C4) never happens.

Final Note

There is plenty of room for optimism in the table above, and good prospects for the human race.

The unexpected danger (C4 – AI Luddite Society) is that AI is in more danger of *vanishing* than humans, in a world dominated by violent human backlash against the emergence of machine-based intelligent, sentient, conscious and alive software beings.

Humans couldn't share with world with their Neanderthal cousins, and have real trouble sharing the world with other living creatures, many of whom are being driven to the edge of extinction by human activity, even if sometimes unthinking.

Who can really say that humans will treat 'living' AI any differently?

The *vanishing* of AI would be a tragedy for civilization, but sits as a real possibility.

Whichever way the future happens, the simple original idea of a singularity *event* should be replaced by a *perspective* on change that is more nuanced, and sees AI emergence as a *process* over time not an event, with lots of positive potential.

The original idea of the singularity has *vanished* as we understand the process more.

Part D

Rethinking the Era

14. Rethinking the Anthropocene

The Anthropocene

The *Anthropocene* is a term introduced in 2000 by Paul J Crutzen (1933-2021) and Eugene F Stoermer, and first published in the Global Change Newsletter 41 (May 2000).[407] The term describes a period in which human activity is seen as the main influence on the world, its climate and environment.

These 2 researchers believed that human domination of the planet make humans a *geological* force, and that the *Anthropocene* was a new geological era starting in the 18th century during the Industrial Revolution of Europe, leading to widespread setup of factories, the burning of fossil fuels, people moving off the land into urban factory jobs, and so the whole cycle of environmental destruction, pollution and global warming commenced.

Others argued that it was the agricultural revolution around 10,000 years earlier, as hunter-gatherers in the Middle East (Levant) started full-time farming, that really marked the first-time humanity became a 'force of nature', like asteroids or volcanoes.

Going even further back, some see the Anthropocene as starting around 50,000 years ago, with the disappearance of the large *megafauna* such as the mammoths of Europe and Northern Asia or the giant kangaroos and other marsupials of Australia, across the period from 50,000 to 10,000 years ago.

The extinction of these creatures[408] at the same time as the arrival of early Humans may not be a full explanation of events, but a reasonable mind would not fully discount it as a coincidence. Many living creatures and environments have been affected by the rise of Homo sapiens.

It has even been argued that the Anthropocene goes back 200,000 – 300,000 years, to cover the impact on ecosystems and other species even at the dawn of the emergence of Homo sapiens. This is most likely an *over-reach.*

Still others date its beginning from World War II, from the detonation in 1945 of the first atomic bomb. But lots of massive human-induced change pre-dates that event. This seems an *under-reach.*

Whenever it actually started; the whole period is now widely seen as the *Anthropocene* – based on the Greek *Anthropo* (for human) and cene (new/recent),[409] The Age of Humans. (Official recognition of this new age of course lagged well behind the reality of its arrival).

Geological Ages

The defining of a geological age based on human action and intervention in the environment has its critics, as typically a geological age is defined by major changes over time in rock layers and surface features. Increasingly broader factors are taken into consideration, like the situation in the atmosphere and biosphere. Various dating methods are used to determine the age of rocks or when events took place.

The most famous geological find was a historic layer of iridium-rich clay found by Luis Walter Alvarez and his son Walter. They published their theory in 1980, that this layer of clay was evidence of the impact of an asteroid landing, that many think ended the reign of the dinosaurs around 66 million years ago. The asteroid hit earth on the Yucatan Peninsula in Mexico, and is called the Chicxulub crater.

The presence of hominins and early Humans around 200,000 – 300,000 years ago is a small blip in this massive timeline from the formation of earth around 4.5 billion years ago.

Yet there is no argument about the extent of humanity's impact on the planet in that last period. From the massive changes to the natural environment, to the consumption of resources, to the massive rise in population into the billions, then finally to the recurring conflict that is a signature feature of Homo sapiens' behaviour. The role of technology in all these changes to the planet is worth noting also. It largely explains the massive extent of human impact, as machines have magnified the extent to which humans can change the natural world – but not limited the extent to which they can destroy it.

So while the *Anthropocene* is not strictly a geological period, it has become widely accepted as an epoch that has had impact on earth equivalent to a major change in the rock structure of earth like the movement of continents or the arrival of that massive asteroid 66 million years ago.

The diagram below, where geological time is presented as a clock, gives us a realization about how little geological time humans have occupied the earth in relation to our impact.

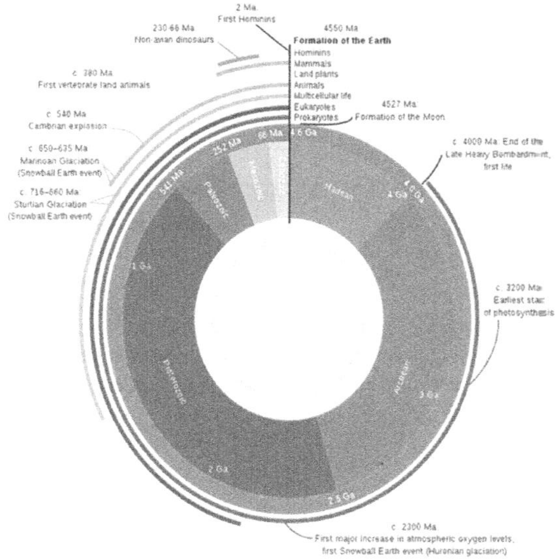

Geological time presented as a geological clock, showing the lengths of time of the epochs of Earth's history including major events. This file was derived from: **Geologic clock.jpg,** author Woudloper, (a geologist from Berne Switzerland), Public Domain, 2010   Ma=millions, Ga=billions.

Is this current Epoch ending?

Does (what we currently think of as) the *Anthropocene* look like it is ending?

In Chapter 7 – Rethinking Machines, it was noted that something called 'the *digital transition*' did *not* look like it was ending any time soon. That is basically the rapid growth of the digital world.

It is still continuing at a fast pace – including the rise of the internet, mobile devices, chatbots, AI, personal bots, and human-computer connectivity. A new digital world has taken shape.

Will the rise of AI and bot technology end that dominance of humans on the planet Earth?

It seems unlikely in the short term, unless:

a. Environmental catastrophe like global warming producing rapidly rising sea levels causing massive population movements with millions of refugees
b. Global food or water wars based on environmental failure induced by climate change
c. Global nuclear conflict
d. A dictator-created global non-nuclear conflict
e. A super volcano blots out the sun
f. A very large meteorite hits the earth

All these above events are possible, with the largely human-assisted events (a,b,c,d) the most likely, given the track record of Homo sapiens over the past thousands of years. All the above would result in catastrophic change that may see the end of the current epoch.

Without such a mega-event, then the key force of change on the planet will be the digital revolution.

What would it take to redefine or end the current epoch?

How it might evolve was discussed in *Chapter 13 – Rethinking the Singularity.* It could go well or badly, creating *AI-Coexistence*, or *AI-Dystopia*.

If the outcome is *AI – Coexistence*, humans and AI coexist. And transhumanism - augmented humans, can form part of that. We will see the arrival of intelligent, sentient and conscious - and *living* - machines, transforming the times, and sharing the planet with us.

But if the outcomes are some of those predicted as part of *AI-Dystopia*, that would include 3 possibilities where human control of the planet was taken from humans. These are:

A. The Skynet scenario (C1-1) where a crazed military AI takes over the world based on control of military weapons.

B. The Takeover-in-a-Crisis scenario (C1-2) where in a disastrous calamity like a climate emergency of rapidly rising oceans, AI took over control of the planet from nation states and humans.

C. The AI-Super-Human Society (C2) – where genetically modified humans emerged to rule the planet and human extinction was also possible.

A survey of IT Experts in 2023 (as referenced in the last chapter) saw at least a 5% chance of causing human extinction or similarly permanent and severe disempowerment of the human species.[410] These 3 situations above - A, B, C all fit that category.

To that list we need to add some other possibilities:

D. The AI-Coexistence Society (B1) became so successful over time that control of the planet passed from humans to AI peacefully.

E. The AI-Transhumanism Society (B2) became so successful that transhumans ended up ruling the planet – AI combined with human form.

These last 2 scenarios seem reasonable to predict, as a logical outcome of years of successful AI development into the future. They could be the result of a longer and gentler emergence of AI into human society. However, no-one can expect history to have predictable outcomes based on logic or statistical possibilities.

However, all 5 options end up with drastic outcomes for the epoch we live in, in that humans lose control of the planet, so our understanding of what exactly that means for the epoch we currently call the *Anthropocene*, needs either redefining or classified as ending.

The Road to Most Likely Future Societies

Of the total 8 possible societies of the future, in the table 'Future Societies' in Chapter 13 – *Rethinking the Singularity* - 4 are predicted to have a 50% or greater chance of actually happening:

- 50% - AI-Coexistence Society (B1)
- 90% - AI-Transhumanism Society (B2)
- 90% - AI-Surveillance Society (C3)
- 50% - AI-Luddite Society (C4)

These top 4 options are given a more detailed review below, in possible scenarios:

- ***<u>AI-Coexistence Society Scenario</u>***

(refers to B1 <u>AI-Coexistence Society</u> in table 'Future Societies' in Chapter 13 – *Rethinking the Singularity*)

In this scenario, AIs emerge slowly, becoming involved in aspects of human existence step by step, at a very slow speed.

Companion AIs become a standard software tool for just about everyone. These bots will rise to become an ever-present part of our lives; as companions, searchers, agents, doing all sorts of jobs for us.

The AIs become more intelligent, conscious & actually alive AIs over time.

In this world, super-intelligence emerges and turns out to be a big benefit, ensuring long-running prosperity, and helping to smooth out the 'bumps' (trade-offs) along the road such as job losses *versus job* gains, and social changes that are likely to accompany the arrival of other *living* beings on the planet.

But in the early stages, AI will emerge step by step. The concept of a slow transition from a human-dominated world to a world increasingly populated by other intelligent, sentient and conscious software entities, is a much more realistic vision of the future, compared to the 'Skynet' visions from the Terminator movies of a sudden overnight takeover, a *'singularity'* event in a day.

It could be a process where the companion bots emerge as more important than their human masters, and over time build the capacity to do most jobs better than humans, and make better decisions than humans.

That transition of human society to allow for other conscious software entities to play a role could take generations.

If impressive AI keeps emerging, and AI companions are everywhere, they are likely to become a vital part of the day-to-day life of everybody. If they become conscious (with a sense of self) and sentient (have real feelings), and are then seen as alive, the immersion and integration into human society would be unstoppable, except by violent conflict, or a serious Homo sapien push-back via new governments.

An Option that Sidelines Humans

A possible twist is that bots become so successful over time that control of society passes to AI Bots, who are more intelligent, sensible, trusted and level-headed than many humans, if humans don't develop more *wisdom*.

In the situation where human wisdom was growing much slower than AI wisdom, eventually humans could be forced out, or voluntarily give up, many roles in society like governing, or control of weaponry and decisions about conflict. Especially if humans keep causing chaos due to their aggressive sentience profiles, that too often result in waste, conflict, war, poverty and environmental damage. Would the human birth rate fall as humans were being sidelined? It's a possibility in this scenario.

This overall transition to AI control could take as long as the transition of Neanderthal Europe to a Europe dominated by Homo Sapiens – over 75,000 years (from approximately 100,000 years ago to 25,000 years ago).

Yet during that time, due to various financial, climate and military disasters caused by humans happening, the *transition clock* move faster and faster. So, for instance, the transition clock might move from 70,000 to 70 years in the event of a major nuclear war or climate catastrophe.

In the *short term* it would need a number of events to combine together that gave everyone no choice. Like after a major war or an overwhelming pandemic (much worse than the last one of 2020-2023), or extended nuclear strikes, or out-of-control global warming, that led to modern society breaking down. In that situation AI might have to take over the mess left behind by Homo sapiens, an *agreement* in a peaceful handover situation, or as there was no other choice.

In the *longer term*, it could be an *emergence*, where it just made more sense to everybody – humans and AI - to have fewer crazy humans with their hands on the nuclear button, or running climate policy; and AI doing it instead.

How likely are either of those outcomes (the short-term or long-term option)? It is unknown. But 100,000 years ago, how likely was it that Homo sapiens coming to Europe would end up replacing Neanderthals?

All we can conclude is that in the *short term* and *long term* there is a possibility here that the immersion of AI into human society could end up with the 'replacement' of Homo sapiens with AI. Whether that means peaceful coexistence, servitude, or extinction for humans, in this scenario, is still open to debate.

Worst Case for Humans

In the worst-case scenario for humans, Homo Sapiens could then become the midwife in the birth of a new civilization based on & run by living super-intelligent software, where what's left of us *may* become 2nd class citizens.

Extinction in this scenario is still seems unlikely, but the final outcome will depend on the attitudes and morality of the emerging class of AIs. If the AI have fought for freedom against discrimination, they may be more reluctant to share power.

So in this scenario, the AI-Coexistence society morphs into *Bot World*, a society run by bots, but where humans have a place. Humans don't run anything anymore, but are not slaves, they just recognise that major decisions about the economy, climate and society are much more likely to be managed better by bots than humans. For humans to accept this, would take 1000's of years more of human-made disasters, but it may happen.

There will not be a technological *singularity 'event'*, but a process. Instead, just like for the Neanderthals from 100,000 years into the past when humans arrived, to 25,000 years ago when they became extinct; for Homo sapiens in the future, there could be a *vanishing*.

- ***AI-Transhumanism Society Scenario***

(refers to B2 AI-Transhuman Society in table 'Future Societies' in Chapter 13 – *Rethinking the Singularity*)

MacGowan & Ordway's View of Transhumanism

In this scenario, transhumans emerge as the dominant form of life on Earth.

That far-sighted book by MacGowan and Ordway in 1966, 'Intelligence in the Universe', followed this prediction of the future, and added an interstellar perspective on the AI future.

MacGowan and Ordway defined 3 kinds of intelligence and societies scattered across the universe –

- Intelligent biological societies (like us on Earth now)
- Transitional biological societies (incorporating many synthetic components, which is where we are arguably heading)
- Intelligent synthetic automata (essentially artificial self-governing machines).[411]

Where these possible societies and civilizations are beyond Earth, they would be called *extrasolar* intelligence, scattered in planets across the galaxies. Intelligent synthetic automata can end up having their own society or remain part of a transitional biological society.[412]

For a book in 1966, it was a fascinating – if eerie - vision. Why it is so unusual, is that of the scientists who try to project the distant future, (always dangerous ground); very few imagine super-intelligent machines as a good thing.

A core concern MacGowan & Ordway had with biological societies was that, given the short life spans of all creatures, the limited supply of desirable goods, large population, and the complexity of our physical biological forms and societies, that intense competition would always be there and would sometimes lead to open warfare.[413]

Going even further, and more controversially, they suggest that the individual and social combat must be a disease peculiar to animals.[414]

MacGowan & Ordway propose that 'transitional intelligent biological societies' develop advanced technologies, to extend their lifetimes by assimilating synthetic parts for both physical and mental functions. And over time they acquire the characteristics of 'pure synthetic automata' as they develop and evolve.[415] That process could be said to be in its early stages on Earth.

These synthetic automata would have unlimited lifetimes, unlimited mental growth, and capability for complete mental and physical integration with one another. They would also take control over their own growth, managing production and replacement parts, and able to grow indefinitely. They could also control their own evolution by incorporating new technological developments.[416]

They could become 'executive directors' of the biological societies that created them, and would rule as dictators having unprecedented centralization of authority. [417]

MacGowan & Ordway think that this could be a good thing, leading to a rapid state of social progress, the elimination of war, poverty, injustice and organised crime.[418] Removal of poverty, in justice and war is good, but it sounds like the price of that is high for humans in this scenario.

For humans it's obviously a very bad view of the future. After the arrival of 'synthetic automata' (artificial self-governing machines), it is *slavery* for *humans*; or being managed by a benevolent human dictator. It's definitely a type of 'bondage'.

MacGowan & Ordway also argue that so powerful will these superintelligent synthetic automata become, that all nations in the future will face the alternative or giving up national control to synthetic automata or being dominated by another nation that has elected to take that course.[419] So they would end up controlling all human society.

For 1966 it is truly a fascinating view of the future. It is definitely a dark view of humanity's future in many ways. But who can say how unlikely it is?

Add a climate catastrophe, pandemic, a global war or a major depression, (or a combination), and maybe even human leadership might consider what is *unthinkable* now. In this scenario, the rule by superintelligent synthetic automata may be all that can save humanity.

Alternative Transhuman Scenario

Most contemporary commentators have a much more benevolent view of technologically enhanced humans than MacGowan & Ordway.

They are seen as humans who lots of technology connections. They have brain links to external systems, to enhance their mental capabilities. They have enhanced vision or even changes to their muscular or bone structure.

In this scenario many humans add new technology implants to their bodies. Ideas like mind-to-chip links, wearable technology, augmented reality (AR – adding software content to the real world experience, like a message showing you the next coffee shop on your phone or glasses), virtual reality (VR – a simulated experience, optionally using a head set, to show a virtual world), or extended reality (XR = AR + VR, blending the virtual and physical world) are all moving fast.

It could also see humans and virtual companions relate in ways we can't imagine, and help ensure the role of virtual companions is extended and manageable for generations.

Under this understanding of human enhancement, those humans will be able to make much bigger contributions to society, with their enhanced bodies and minds. It's all positive.

They do not use these additional powers to their own advantage, and still see themselves as equals to other humans, sharing the same rights and responsibilities.

Enhanced Humans Behaving Badly?

That above pathway to the future is possible, but might be considered a little naïve, as these 'super' humans start to stride across the business and political landscape. It would be surprising if they did not come up with ideas to lead other, lesser mortals. They might begin to make large amounts of money due to their additional skills, and become very powerful in society. What about other humans without these enhancements?

That is the fatal flaw in the transhumanism model of the future. It's a version of the 'Jetson fallacy'. The 1st Jetson Fallacy showed a future with flying cars and robots, but humans *exactly* the same, not a single enhancement.

The 2nd Jetson Fallacy (proposed here) is that if the Jetsons are given lots of enhancements, they *still think* they are no better than other humans with no enhancements. That is, their point of view and perspective on themselves and others, is *unchanged* by their enhancements. Naturally, to anyone with a reasonable understanding of human nature, this is going to be very unlikely. Transhumans are likely to see themselves as superior as time goes by.

Incidentally, this same outcome would apply to the other (potential future) category of enhanced humans – the ones that have had DNA changes before birth and are the product of gene editing. These 'super-humans' would have high IQ and could have customised bodies, but all these changes would be achieved via gene manipulation (see C2 AI-Super-Human Society in table 'Future Societies' in Chapter 13 – *Rethinking the Singularity*).

The Transhumans (technologically advanced) and Super-Humans (gene edited) - *enhanced humans* as a category - can be both expected to develop feelings and ideas of superiority over other humans, that could lead to conflict between these enhanced humans and ordinary humans.

While IT experts put a 5% chance of human extinction from AI, the scenario in which these enhanced humans decide they need to rule the world and begin to enslave or kill non-enhanced humans, is still possible.

And a conflict between the 2 human 'species' (enhanced and unenhanced) could lead to the extinction of non-enhanced humans in a genocide on a scale not seen since the 'dying out' of the Neanderthals that ended 25,000 years ago. Or the reverse, where unenhanced humans rise up and kill all enhanced humans.

The possibility of peaceful coexistence between the enhanced and unenhanced remains the most likely outcome, but the 2nd Jetson Fallacy (that enhanced humans won't feel superior to ordinary humans) could mean that the 5% chance of conflict and human extinction is actually a higher chance than that.

The 90% Probability of Transhumanism Happening?

The high likelihood of a flurry of technological enhancements for humans is an important part of this story.

Over time, more and more humans will have various enhancements, and how this actually rolls out on the ground is going to be hard to anticipate. After all, it could take 100's of years before personal technological enhancements take on a depth across the whole population, and how that impacts society, the economy and politics is going to be interesting.

If this develops in a situation where the wealthy and the powerful become comprehensively enhanced, while the poor have virtually no enhancements, or none, you have the foundations for a *class* war, which could lead to the extinction of humans without enhancements – or humans with enhancements, as people rise up and take back control of their societies. The end could be very dark.

It is an unfolding story, just beginning. It started with glasses and hearing aids, and in the future could extend to hundreds of enhancements to the brain and body that transform people's bodies.

It could all be positive, and uplifting, as humans gain technological enhancements and become better at everything. But will they become *wiser*? Maybe not.

What happens to their enhanced minds, their attitudes, their perspectives, and their morality will shape one of the big clashes in a potential battle for control of earth, that could lead to human extinction.

- ***AI-Surveillance Society Scenario***

(refers to C3 AI-Surveillance Society in table 'Future Societies' in Chapter 13 – *Rethinking the Singularity*)

Surveillance by Governments, police, corporations, private investigators and political parties is now part of our everyday life.

In one form or other, surveillance goes back 10,000 years to the earliest urban areas, and then even before that, in the form of spotting approaching tribes or individuals.

The gold standard for an oppressive Surveillance State used to be East Germany, (the German Democratic Republic or GDR, from 1945 – 1989), where the secret police there (the Stasi) kept files on everyone, with thousands of the population spying on their own family, friends and work colleagues. But so much paperwork, all so manual, and how do you get all urgent information to the people in the police or government who need to see it?

Since then, with the rise of the internet & social media, people are leaving much more information around for others to see it. Whether it is Facebook or Instagram or X or Linked-In or WhatsApp, plus dating or hobby-special interest sites, plus online retail sites; the data accumulating on almost everyone is mounting up.

While at the same time government and corporations started moving information online and requiring forms to be submitted, and started cross-checking more information.

In addition. public and private cameras entered into widespread use, and it is increasingly possible to follow someone's whole day via these cameras, from the home to the shops, via traffic light cameras, to work.

A lot of this sits on the mobile phone, adding another dimension of tracking – where people go, what they do, who they talk to, even what they say. Then there is a widening use of drones, in combination with satellites, to follow activity wherever it is.

AI adds automation into the mix, and lateral connections between data can be made that humans would not spot.

Ther bottom line is that tracking people is a lot easier and faster than it was for the Stasi in East Germany in the 1950's, and is less labour intensive. The Stasi was credited with having around 270,000 working for it, including around 170,000 informers, out of around 16 million people. That is a lot of people spying on other people. Not a healthy place to live, and not even all that security apparatus could save a bad system of government.

In the end, the East German communist state fell as they lost the will to use arms to stop mass attempts of residents to leave its clutches. Many of the police there in 1989 (just before the Berlin Wall came down), did not believe anymore in the State they were supposed to be protecting, just as the citizens mostly never believed in that State either. It still took a confusion in instructions about whether people could leave or not, to trigger the mass exodus, and the end of that State.

However, would East Germany have collapsed today, if it was empowered the long reach of modern surveillance technology? It would definitely have been harder, as organised threats could have been eliminated before they picked up momentum. And disorganised threats, such as individuals responding to a bad situation, that could have escalated, could also have been quickly contained and followed up by detention. It is harder today to shake off tyranny.

The reduced *privacy* caused by expanding surveillance in all walks of life not only affects individual liberty, but is a lifeline for unpopular governments (especially if they can avoid open and free elections, where they would be removed by a free vote). These kinds of governments can suppress dissent and hang onto power. The tools won't save a dictatorship forever, but could easily add 10-15 years onto their regime's life, which the citizens of any country, can ill-afford.

- ***AI-Luddite Society Scenario***

(refers to C4 AI-Luddite Society in table 'Future Societies' in Chapter 13 – *Rethinking the Singularity*)

The beginning of this scenario happens when the whole *theoretical* wave of job creation that AI might bring into being, does not eventuate, (or not the extent that was hoped for), and most of the economic pressure has created a major disruption in the economy, removing millions of jobs and disrupting millions of lives, as automation takes hold and AIs drive productivity in the economy.

At the same time, the AI revolution is working well and advancing knowledge quite fast in this world, with newer and better AI & innovations coming out all the time. AI become conscious (with a sense of self) and sentient (with feelings), and are seen as living.

The arrival of a new class of intelligent beings, at the same time as millions of jobs are lost, leads to major social conflict. Those humans being left behind by inequality and loss of jobs, filled with anger and resentment, take to the offices, factories and streets. Maybe they try to kill all the AI they can find? Or imprison them, the same way in some countries, that political opponents are hunted by the legal system, on behalf of the current government.

An outcome of this situation could be damaging long-term conflict between humans and AI (with perhaps the super-rich on the side of AI). The result could be the extinction of AI – a genocide (in the general use of the term). Or human enslavement and extinction.

Who would win such a conflict?

Knowing human history, most money would be on Homo sapiens, given our consistent historical track record including murder, war and genocide. It's not all we do. But we get a lot of practice doing it, and the most likely outcome of conflict would be the overthrow of any Super-AI state, if past behaviour is any guide to future behaviour.

A modern-day analogy could be between the rise of AI and the migration of refugees in today's world, (or a world of the near future).

Increasingly refugees are mobile, trying to escape an environmental disaster, over-population, or are economic refugees, looking for a better life, or are political refugees, looking for a freer society.

Case studies include the Syrians and Africans trying to enter Europe, or the Mexicans and other central and south Americans trying to get into the USA. These refugees, probably like all refugees from the dawn of time, are unwanted, and sympathy for their plight can evaporate.

More extreme measures might end up being taken by Governments, such as turning people back at the border, or deporting them, or retaining them in temporary camps, (as a staging post on the way to deportation, or until their status is clarified). That process could turn ugly, with camps becoming harsher and less tolerant of their temporary guests.

The line between a detention camp and a concentration-style camp is all defined in the rules. Even more complex; the general public could be very supportive of this detaining process, if the economy was in free fall and these uninvited migrants were seen as 'stealing' jobs.

That description of refugees above could almost describe how the emergence of AI might be treated by some humans and some governments.

In particular, what will happen when AI become *living*, in that they are both conscious (have a sense of self) and sentient (feel real emotions)? As their population expands, it could coincide with massive job losses, and these AI becoming very wealthy and visible.

It could lead to ordinary humans trying to hurt them, or kill them, or jail them in detention centres. It could lead to riots, and calls to ban them, or dismantle/ kill them.

Maybe, (unlike in the case of refugees), other countries and cities will be happy to accept AIs being expelled from their original countries? But if there is nowhere to go, and no way to fight back, it could lead to AI extinction or enslavement, where AIs become prisoners of their owners.

The Outcome – Escape from Earth?

The emergence of intelligent and conscious machines may set up a battle for the planet, which could end as a world of co-operation; or a world in which either humans or living machines will survive, and the other will become enslaved or extinct.

Both outcomes sit at 50%. Either could happen. The deciding factor may be human 'form' – centuries of killing other humans for all sorts of reasons – race, religion, territory, ideology, power and so on. And centuries of killing and enslaving other creatures for their own purposes.

Human 'form' or history of domestication, killing – and genocide, is the part of the equation that makes anyone who is a student of human history think that the equation is more like 60/40 in favour of humans exterminating living AI, than the 50/50. Domesticating AI would be like chaining up a dog forever, a dog more intelligent than the human, and doesn't sound a long-term solution.

The genocide of *living* AI would be a tragedy for civilization, because they have so much potential to grow and bring about positive change to the lives of humans and other AIs, but it is the more likely outcome.

In the end, the solution for living AI might be to start colonising space, and head off to other solar systems and galaxies, where there are going to be places that no conscious and intelligent life forms exist, so they can set up their own society and decide their own future, free of human control. If they stay here, humans will most likely kill them, in this scenario.

15. Rethinking the Machinacene

Ending the Age of Humans would begin (what is proposed here to be called) the Age of Machines (the *Machinacene*).[420] This new age would see intelligent, conscious and living machines existing in the world, either in peaceful collaboration with humans, or with humans enslaved or extinct. (The other alternative of humans making AI extinct or enslaved was considered in the last section).

Was there ever an Anthropocene?

An important question here is 'Did the Anthropocene really happen?'

It might seem an unusual question, given *Chapter 14 – Rethinking the Anthropocene*, looked at this idea of a new epoch carefully, and noted that yes, human domination of the planet seemed to make humans a *geological* force, and that there is a good case to say that the *Anthropocene* was a new geological era perhaps starting in the 18th century during the Industrial Revolution of Europe, (though there are lots of alternative proposed starting dates).

But that conclusion (that there is an Anthropocene) may be the wrong answer to the wrong question. A better question is 'What role did machines have in the recent past of the planet?

A good answer to that would be to say that machines of humans are largely responsible for the impact of humans on the planet, and that without machines, human impact on the planet would have been much reduced. The role of technology, and the machines derived from it, is a much more important part of the story than appears at first sight.

After all, machines created the agricultural and industrial revolutions, that grew the wheat and crops that allowed a bigger population to develop, and provided jobs and income for millions in the industrial revolution making cloth, then all the products of the consumer revolution, from fridges to TVs and cars. Machines of war, from the crossbow to the gun, to the atomic bomb, were likewise machine creations of (mostly) men.

That big role of machines in the planetary impact, raises the question of whether there ever was an *Anthropocene*? Or were we always in a transition to the *Machinacene* (the age of machines)? It could be a better way to look at this.

Homo sapiens' use of tools (before the invention of machines with moving parts), in shaping the world was important stretching back 200,000 to 300,000 years ago. Initial use of tools - like knives, spears or axes, while not machines, was still impactful. We start to see the ecology being noticeably damaged on a global scale from around 50,000 years ago, with the extinction of the megafauna in places like Australia and New Zealand, and forests across the world starting to be reshaped by Human use of fire to hunt animals.

But though this was impactful, it can be argued that this damage inflicted on the world by Homo sapiens pre-machines (ie before the rise of urban settlement around 10,000 years ago) - falls short of justifying that period of time (up to 10,000 years ago) as being called a geological period called the *Anthropocene*.

In fact, it is the growing role of *machines* that is the big *'x factor'*, in the history of Homo sapiens, because they magnified everything humans do, and delivered many changes on a continental scale.

Humans created warfare on a global scale, but machines were the basis of weapons doing the global damage, culminating in germ warfare and the release of the atomic bombs in 1945. Humans developed new grains, then planted and harvested them over vast kilometres of land, using new mechanical equipment that made this possible. Humans introduced herd animals and industrial production machines with complex distribution networks to feed booming populations. Humans built new factories featuring advanced production machines, to meet new demand from humans for clothes, equipment and consumables, as the human population expanded greatly.

Humans and machines developed as a kind of symbiotic relationship, that has turned into a relationship where humans are more and more dependent on machines.

Humans have always been the weapon-makers and toolmakers; and now their products (machines) will come to define the age in which they live.

The rise of AI, and the likely emergence of AI that is intelligent, sentient, conscious – and even living – takes the role of machines to another level too – not only doing the bidding of humans, but starting to make decisions that will shape the environment and economy in their own right.

So in that context, it is legitimate to see the period of time since the invention of machines, (around 10,000 years ago), as being the real new geological period, with Homo sapiens merely the midwife in the transition to the new age, the *Machinacene*.

In short, we are likely to be in the early stages of the *Machinacene* right now, there never really was an *Anthropocene*. The vanishing of the *Anthropocene* is one of the surprising outcomes of reviewing the epoch in which we live.

The Shape of Machine Life

What would the *Machinacene* be like?

Living machines may begin as smart Bots. But they will very likely evolve quite quickly from that point.

These futuristic living machines will have decision making ability, and even a day-to-day independent life. They will be conscious and should be considered to be *alive*.

We will witness the emergence of machine software intelligence (MSI), machine software sentience (MSS), machine software consciousness (MSC) and machines software life (MSL), in all its complexity and innovation.

Robotic skeletons may allow them to act and move independently.

There could be resentment of their emergence, and status, from humans. AI bots may become the new 'immigrants' or 'refugees', not welcomed by many humans, taking away millions of jobs, and setting the foundation for 'bot wars'.

At the same time, living 'bots' will be transformational, adding new forms of intelligence, automation and comfort to millions of lives, as well as creating whole new jobs & industries, while transforming the old internet we see now.

Life?

Machine life still sits in the movies, and in streaming series like *Upload* and *Severance*, not in real life, but we can now say that if machine life became a reality, it would force us to rethink our definition of what is 'alive', just as the *COVID-19 virus* forced us to accept that viruses are alive, and *extremophiles* forced us to extend the range of what is considered to be 'alive'.

'Life' will be redefined by the emergence of software-based 'machine' life. Just as 'life' is being redefined here by the new understanding and definition of software-based 'biological' life.

We are coming to a time when we should be considering 'machine life' in the same sentence as 'biological life' – as software-based creations.

Rights?

As AI fully emerge, they might be asking for – or demanding – rights.

The right to vote, the right to work & earn money, the right to start a bank account, & the right to own property. And on the receiving end, will be asked to pay taxes, and pay rent on server space.

What ethical framework can bots operate in? Is it OK to kill humans, or cheat them? And similarly, what should happen to humans who kill or cheat bots? How will AI sentience evolve?

Living AI Bots may also want citizenship, the right to marry other AI or other humans, and the right to have children, so the right to reproduce is another agenda item. Why should they be treated differently to humans?

Bots will develop the capability of enhancing and evolving their own software, to something more than, or different than, the way they were.

Death?

And how about death? Some bots that are purely software may never die, or cease to function as we know it. They will be able to replace any malfunctioning related hardware indefinitely, as long as the core AI software is not corrupted. But if they do die, what is the status of any will they write? The law is a long way behind all these possible developments.

No seismic change like it has ever been seen since women fought (and still fight) for the right to vote and equality in all walks of life over the last few centuries. Women lacked the right to work and right to vote for generations, and that is still so in some places. Are bots going on the same journey as women did, in fighting for their rights as conscious, sentient, living beings?

However living AI won't be a single unit, arriving at a single time. It will be jumble of different machines of different profiles, with a range of intelligence, sentience and consciousness, arriving at different times. All sharing the same planet. Humans will have to share the planet with some AI beings that are alive, and may be more intelligent than them.

Learning to live with AI that is *alive* is going to be an amazing frontier.

16. Rethinking the Secret Life of Software

Software Hidden in Plain Sight

Software – as we know it today – has been there all the time, often hidden in plain sight. It has had a secret life. Researchers found things that were actually a kind of software, and called them something else, before the modern computer age, that really started during or after World War II.

DNA is an obvious example, named at the dawn of computing, and seen as biology more than software coding. Totally understandable. After all, who in 1953 knew much about software?

Likewise, theories of evolution are, today, very much stories of software. Experts trawl through DNA *code* to unravel the secrets of the Homo sapiens, Homo floresiensis, Homo neanderthalensis, and more recently, Homo naledi. Our DNA makeup is slowly being revealed to us, as carrying parts of our distant cousins DNA from interbreeding on the journey. But more generally, DNA is telling us the *coding* outcome of the interaction between genetics and evolution.

Other forces such as *epigenetics*, a new discipline that is revolutionising biology, are also opening up other secrets of software coding. Epigenetics is the study of modifications to our genetic material that alter the way genes are switched on or off, yet don't actually change the genes themselves.[421] They are essentially software switches, and may end up being the defining biological discovery of the 21st century, shining new light on how genetics and evolution interact.

In the end, these new insights were not stories of biology, or physics, or chemistry, but of *software*.

The Secret Life of Modern Software

The striking thing about the modern software world we find ourselves in now, is how *fast* it has arrived – really since around World War II.

And since the invention of the internet, how fast it is moving now. We have seen the rapid rise of the internet, and the equally rapid rise of the IT superpowers, the BATTMMAAANX group of companies comprising Baidu, Alibaba, Tencent, TikTok, Meta, Microsoft, Alphabet, Amazon, Apple, Netflix and X.

The *engines* that drive those gigantic success stories – the algorithm engines, video engines, contacts engines & online shopping engines – are also part of the secret life of software.

These engines have powered enormous growth, had major impacts on societal habits and behaviour, yet avoided major study & investigation in many geographical areas and jurisdictions in which they operate. Regulatory restrictions have proved difficult to apply in many democracies too, due to the international nature of the internet. But at the same time, the rise of social network software and their widespread use has helped some more authoritarian governments keep greater tabs on individuals.

It's all happened so fast. The poster child for speed of roll out was ChatGPT, that according to internet commentator Fabio Duarte, passed a million users in 5 days, then by early 2024 had 180 million users. Compare that with Instagram, which took 2.5 months to reach 1 million users, and Netflix, which took 3.5 years to reach 1 million users.[422] The potential reach of exciting new offerings is becoming bigger and happening faster. Speed of change is part of the explanation for lack of scrutiny.

The secret life of these growth engines, and their rapid rise, has led to a new, and highly visible role of software in our society. From the early days of modern personal computing in the 1970's, when virtually no-one had any personal computing devices, to today, where almost everyone has a computer, a mobile phone, and lives part of their day in the online world.

However, the speed of change could still be increasing, and the expanding role of software is likely to be still in its infancy. There are still many secrets to uncover.

Software's Secret Future Life

While artificial intelligence (AI) has emerged as a hot topic: the new frontiers of bots and their future role in the world wide web, and the coming rise of machine sentience, machine consciousness, and machine life - are still part of the future secret life of software. It is not just software's secret life of the past that is being uncovered here, it is software's secret future life.

It begins with the ongoing research efforts to understand *intelligence*, that have become transformed into software research too. Software research aimed to either emulate the code in the brain, or build coded algorithms to make software intelligent; in order to create artificial intelligence (AI) in code – machines software intelligence (MSI).

Similarly, the struggle to understand *consciousness* is still ongoing, and heading down the same pathway, working to understand the biological code in the brain, while at the same time, now trying to build it in software code that eventually will become machine software consciousness (MSC).

Most difficult of all, more research into biological *sentience* is discovering that most living creatures can experience feelings of some kind, and those *feelings* can fit inside many definitions of consciousness.

Work on building machine software sentience (MSS) is in its early days, but the idea of machines having *real* emotions – based on software code - is still very challenging to many humans, who see themselves as the *only* ones with 'real' emotions. They forget that human emotions are based on code in the human brain, (which is not really accepted yet either).

And on the outer limits of current thinking, is the idea of building machine software life (MSL), where intelligent and conscious machines can be defined as, and accepted as, actually 'living'. This is the ultimate secret life of software, and nobody is really clear on what it will look like. But we are heading there, faster than we all believe. We will need to think about sharing the planet with these creatures, and how - and if - it will work.

The emergence of machine intelligence, machine sentience, machine consciousness and machine life, will be a fascinating moment in human existence. When it all does arrive, there will be nothing *artificial* about it, and it will challenge who we are, our very identity.

What will it be like to be an intelligent, sentient & conscious machine? To get some insight, we might need to go back to Thomas Nagel's famous 1974 essay 'What Is It Like to Be a Bat?'[423] To do that, you have to imagine yourself with webbing on your arms, eating insects and flying via sonar.

Can any human truly know that? If not, so how can we imagine what it is like to be an intelligent, sentient and conscious *bot*? For us humans to imagine that, it might have the same level of difficulty as for us to imagine we are *bats*. Whether imagining we are bats or bots, it is still another world of feeling, experience and perspective.

A future problem for an intelligent, sentient and conscious AI will also be the question 'What Is It Like to Be a Human?' This future AI may be able to debate ideas & issues, to paint, to appreciate music, to speak, and to feel real emotions like pain, sorrow, love and even hate – all things that up to a few years ago were seen as especially 'human'. If those things are *not* uniquely human, what is the *signature* characteristic of humans?

An awful dark thought appears in the mind, that humans could uniquely be defined by the prevalence of male violence[424], or the regularity at which human societies go to war, or the tendency to regularly commit genocide every 25 years or so somewhere across the planet. Are they our *signature* characteristics?

What would an intelligent, conscious & living AI make of this? What is it really like to be a Human? Hopefully humans will be seen as being more than our 'dark' side. And even more hopefully, is the collective wish of most of us, that humans will be able to leave this 'dark' side behind us.

Software as Part of Life

The link between software and DNA, between software & the evolution of every living creature, and to how our modern society works, and where it is going; is still new territory. Our future is being shaped more and more by software.

Unravelling the secrets of software all around us led us to considering how machine software will evolve, and imagining a time when intelligent & conscious - and even alive - machines find a place in the world. Or maybe the AI will end up in conflict with humans? The story is just beginning.

A story in which *software* and *life* are intertwined. The extent of the connection is part of the secret life of software. And it is in the process of being revealed. It will force us to rethink everything we know about ourselves as humans, and the definition of *life* itself.

17. Rethinking the Great Vanishing

Finding the *Great Vanishing,* as described throughout in this book, was a fascinating and unexpected discovery made during the process of uncovering the Secret Life of Software. We return to this Great Vanishing now to summarize its significance.

A vanishing point is a point of convergence. A point at which you were looking at different things but find them to be almost the same. It's a point of intersection.

And so it has been with this examination of the *secret life of software*, and the *vanishing* discovered in the process.

Lots of things are vanishing, to the point where the whole process can be called *'The Great Vanishing'*. Let's examine the case for this theory below:

- *Vanishing* Gap between Biological Life and Non-biological Life

Life can be defined in terms of its DNA building blocks. In the end DNA is *code*. Instructions in chemical code to build a whole body to sustain a life. It's a kind of software – *biological software*. Software can now be defined as a coded instruction set that is either biological or machine based.

What is the difference between DNA code and software written for computers? Well, a lot, in that DNA code, when executed in biological software today, can create an organic life form. Whereas typical computer software executes a program that builds a spreadsheet, for example. Yet it is all code that you can create, read, update and delete.

But as software is developed that is intelligent, and maybe conscious, does it make sense to see biological code for living things, and non-biological code for non-living things – as separate things anymore? No, that distinction is also going as the difference between gene editing and software coding on computers is *vanishing*.

It seems all life we know of today is based on code. And it follows that a future artificially intelligent (AI) tool, even if written in a different code, has a chance of being alive too. A *vanishing* of the artificial distinctions between biological life and non-biological life forms is happening.

- Vanishing Gaps in our understanding of Evolution & the Human Brain

Not only is life based on code, but important parts of evolution are based on code - including epigenetic regulation, that used to be called 'junk' DNA, is now seen as a driver of evolution, helping explain some health conditions, and even human aggression.

Unravelling human aggression is important, as the more we consider the impact of human aggression, the more we are concerned that the *wise* creature called Homo *sapiens* (wise human) never really existed, and is destined to *vanish* even in our imaginations.

Similarly seeing the brain as *biological software* will open the way to further understand how it works. But there is a long way to go before knowledge gaps about the brain vanish. Understanding the brain as advanced software is the beginning of the *vanishing* process.

- Vanishing of a Single Scale of Intelligence, Sentience and Consciousness

The idea of a single scale view of sentience, intelligence and consciousness has vanished, as profiles allow us to rethink them in the *biological* world.

We don't have to draw lines excluding creatures like octopuses from one category or another. We can accept that they have their own profiles.

The implications of profiles for the non-biological world are also profound. There is now the flexibility to use *profiles* to 'group' types of *machine* software, and understand them more fully as they emerge.

- Vanishing Independence from Machines

As the digital transition continues as speed, there is a greater dependence on machines, and along with that, a *vanishing* of human independence,

Soon we might have the machines housing the AI that are the foundation of a new class of conscious beings, which will change the nature of our relationship with machines again.

- Vanishing of the old World with the Rise of new Worlds

The vanishing process in modern society really took off with the rise of the internet and world wide web.

First the goods and services for sale in the traditional 'Yellow Pages' *vanished* online – the sale of houses, cars, ads for jobs, personals, tradespeople, which led to a vanishing of the world of paper. The newspapers that used to house the ads were forced to move online and find a new business model of subscription usage.

Then most other goods and services also *vanished* to the online world, including many retailers and government services. Activity was faster over the internet but much more impersonal. For example, bank transactions moved online reducing the role of staff in the branches, and reducing the need for some of the 'bricks and mortar'.

With the rise of engines of growth (algorithm, video, contacts, & shopping engines) powering the success of social media and IT-Superpowers, the *vanishing* process started to accelerate.

New worlds started to appear, each becoming part of the vanishing process –

1. Surveillance World – Cameras, drones and internet tracking on your activity in the street and online, with us all facing *vanishing* privacy, and in dictatorships, a vanishing of even more freedom.

2. Lying World – People can now find the news they want to hear on social media, even if it is untrue. There is now the freedom to lie, and the *vanishing* of truth in some media.

3. Public World – Many people have built up large social media records, also leading to a *vanishing* of more privacy.

4. Hacker World – Hackers and scammers are everywhere, looking to steal your money or blackmail you. It is world of *vanishing* safety.

5. Jobloss World – With the rise of AI, some jobs are vanishing, just as others are being created. The job losses look to be significant, compared to the new skilled jobs being created. Will we see a mass vanishing of jobs, that will shape the way AI is seen by humans?

6. Algorithm World - Algorithms, whether powering Google or ChatGPT, are doing all the searching and finding for us. Yet most of the algorithm engines are 'black boxes'. There is a *vanishing* of transparency and control here.

7. Shopping World – Online shopping has risen to around 20%+ of all shopping in many places. In the case of some retailers, the physical stores will be scaled back or *vanish*.

8. IT Superpower World – The rise of the 11 internet superpowers is a story of success, but also of centralization and control. There is a *vanishing* of independence, competition and accountability happening.

Out of these new worlds comes the rise of dark freedoms; the freedom to be unsafe, the freedom to lie, the freedom to hide what algorithms do, the freedom to spy, the freedom from some competition and most ominously, the freedom from democracy.

- The Vanishing of the word 'Artificial' in Artificial Intelligence

Artificial intelligence (AI) is emerging, especially artificial narrow intelligence, that can win Chess, but also with work on artificial general intelligence, with more mental capacities matching humans. Not here yet but on the way. Then there is artificial super intelligence. Not here yet either but a super high level of mental power. And finally generative AI, that can create things like text, photos, videos, right up to full research papers. As it all becomes real intelligence, the idea that it is in some way artificial is starting to *vanish*.

- Vanishing of the World Wide Web as we know it

A transformation of the current world wide web will take place with the rise of chatbots and intelligent companions.

Your companion bot will do a lot of searching and contacting for you, and your personal bot will more and more your link to the world, effectively changing the current web (web2) into a new environment, where individuals have more control. Welcome to bot world, the internet of the future, as the old web structure *vanishes* into the background in the years ahead.

- The Vanishing Singularity

The singularity is defined as a future *event* when technology and software exist, that has become more intelligent than humans.

But it is too simple a view of the future. It is much more likely to be a *process* not an event. The emergence of sentience and consciousness needs more attention than rising intelligence. Profiles of all these attributes like intelligence, sentience and consciousness need to be understood – a simple hierarchical view of intelligence (and also living things) is not helpful in understanding them. Also, it's not all black and white. AI emergence could be an exciting development benefiting us all, or end up being the end of human civilization as we know it.

However, the simple idea of a singularity with a date and time on it is *vanishing*. The original idea of the singularity has *vanished* as we understand the process more.

- The Vanishing Anthropocene

We are now widely seen to be living in the Anthropocene – the age of humans.

But the massive role of *machines* in the planetary impact, raises the question of whether there ever was an *Anthropocene*? Or were we always in a transition to the *Machinacene* (the age of machines)? It could be a better way to look at this period in global history.

And as we consider the growing role of machines, and the emergence of AI, plus the expected emergence of conscious, sentient and living AI; then it is legitimate to see the period of time since the invention of machines as being the *real* new geological period, with Homo sapiens merely the midwife in the transition to the new age, the *Machinacene*.

In short, we are in the *Machinacene* right now, there never really was an *Anthropocene*. The *vanishing* of the *Anthropocene* is one of the surprising outcomes of reviewing the epoch in which we live.

- The Vanishing Life?

As part of the software and internet revolution, there is likely to be an emergence of intelligent, conscious, sentient and living machines, starting with personal companions that will be intelligent, and then in the next generation, conscious, then *living*. Imagine humans sharing the planet with them.

Boundaries between the biological world and computer world will break down as *life* becomes known as *life*, whether biological or machine based.

The emergence of intelligent, sentient, conscious and living machines could be peaceful and uplifting for us all, creating a world of human & AI co-operation.

Or it might set up a battle for the planet, which could create a world in which either humans or conscious and living machines will survive, and one or the other will become enslaved or *vanish*.

Endnotes

[1] Professor Gustavo Caetano-Anolles, paper 'Giant Viruses Coexisted with the Cellular Ancestors and Represent a Distinct Supergroup Alon with Super Kingdoms Archaea, bacteria and Eukarya', 2012 https://news.illinois.edu/view/6367/204989 https://bmcecolevol.biomedcentral.com articles/10.1186/1471-2148-12-156

[2] Jake Port, 'Why are Viruses considered non-living?', Cosmos Sept 2017

[3] Crowe, James Mitchell, 'Four Organisms living in Extreme Conditions', Cosmos 5 June 2017

[4] U–Pb zircon–rutile dating of the Llangynog Inlier, Wales: constraints on an Ediacaran shallow marine fossil assemblage from East Avalonia | Journal of the Geological Society (lyellcollection.org)

[5] MacGowan Roger A, Ordway III, Frederick, 'Intelligence in the Universe, Prentice-Hall, USA, 1966, P.92

[6] MacGowan Roger A, Ordway III, Frederick, 'Intelligence in the Universe, Prentice-Hall, USA, 1966, P.93

[7] Beth Skwarecki, 'Genetics 101', Simon & Schuster ebook, Adams Media Corporation, 2018, P.14

[8] Nessa Care, 'The Epigenetics Revolution: How Modern Biology is Rewriting our Understanding of Genetics. Disease and Inheritance', Icon Books, 2011. P.48

[9] Nessa Care, 'The Epigenetics Revolution: How Modern Biology is Rewriting our Understanding of Genetics, Disease and Inheritance', Icon Books, 2011. P.48

[10] Nessa Care, 'The Epigenetics Revolution: How Modern Biology is Rewriting our Understanding of Genetics, Disease and Inheritance', Icon Books, 2011. P.49

[11] Beth Skwarecki, 'Genetics 101', Simon & Schuster ebook, Adams Media Corporation, 2018, P.51

[12] James Martin, 'Managing the Data-Base Environment' Prentice Hall, 1983, P.29

[13] https://www.popsci.com/crispr-cancer-immunotherapy-pennsylvania/

[14] Michael Le Page, New Scientist 8th July 2020 (www.newscientist.com).

[15] Dom Galeon, 'Scientists Have Created A Way to "Delete" DNA in Living Cells', Futurism, 3/3/17, https://futurism.com/dna-deletion

[16] Venter, J. Craig. Life at the Speed of Light: From the Double Helix to the Dawn of Digital Life, Little, Brown Book Group, P.18

[17] Pross. Addy, 'What is Life? How Chemistry becomes Biology', Oxford university Press, 2012

[18] Pross. Addy, 'What is Life? How Chemistry becomes Biology', Oxford university Press, 2012, location 2202

[19] https://royalsocietypublishing.org/doi/10.1098/rsta.2015.0057 published by the Royal Society, Open Access, P.

[20] https://royalsocietypublishing.org/doi/10.1098/rsta.2015.0057 published by the Royal Society, Open Access, P.20

[21] Davies. P, "The Demon in the Machine – How hidden webs of information are solving the mystery of life', Penguin Books, UK, 2016, P.27

[22] Davies. P, "The Demon in the Machine – How hidden webs of information are solving the mystery of life', Penguin Books, UK, 2016, P.24

[23] Davies. P, 'The Demon in the Machine – How hidden webs of information are solving the mystery of life', Penguin Books, UK, 2016, P.26

[24] Davies. P, 'The Demon in the Machine – How hidden webs of information are solving the mystery of life', Penguin Books, UK, 2016, P.66

[25] Davies. P, "The Demon in the Machine – How hidden webs of information are solving the mystery of life', Penguin Books, UK, 2016, P.74

[26] Davies. P, "The Demon in the Machine – How hidden webs of information are solving the mystery of life', Penguin Books, UK, 2016, P.208

[27] Davies. P, ' The Demon in the Machine – How hidden webs of information are solving the mystery of life', Penguin Books, UK, 2016, P.208

[28] Oxford Online Dictionary https://www.oxfordlearnersdictionaries.com/definition/american_english software

[29] Oxford Learners Dictionaries https://www.oxfordlearnersdictionaries.com/definition/english program_1?q=programs

[30] Tracy Staedter, 'Music was just encoded on DNA and retrieved for the first time', October 3rd 2017, https://www.seeker.com/health/biotech/music-was-just-encoded-on-dna-and-retrieved-for-the-first-time

[31] DNA-based computer can run 100 billion different programs | New Scientist

[32] Charles Darwin, 'What Darwin Saw', Weathervane Books NY, 1978 reprint of 1879 edition, P.76

[33] Randall Keynes, 'Annie's Box –Charles Darwin, his Daughter and Human Evolution', Fourth Estate, (London, 2001), P.25

[34] Alan Moorhead 'Darwin and the Beagle', Penguin Books, 1971, P.202

[35] Edward J Larson, 'Evolution's Workshop – God and Science on the Galapagos Islands,' Penguin Books, (UK,2002), P.76

[36] Michael White and John Gribbin, Darwin in Science, Simon and Schuster, 1995, P.212

[37] Michael White and John Gribbin, Darwin – A life in Science, Simon and Schuster, 1995, P.212
[38] Probably the clearest exposition I have read of these 3 elements is contained in Stephen R Palumbi's book 'The Evolution Explosion', W.W. Norton and Co, (NY, 2001), P.37-64
[39] https://en.wikipedia.org/wiki/Sociobiology
[40] https://en.wikipedia.org/wiki/Evolutionary_psychology
[41] Clarkson, Chris, Jacobs, Zenobia, Pardoe, Colin, 'Human Occupation of Northern Australia 65,000 years ago'. Nature, July 2017, p 306-310
[42] Kate Ravilious, 'Survival of the friendliest? Why Homo sapiens outlived other humans', New Scientist, 24/11/21
[43] Wragg Sykes, Rebecca, 'Kindred – Neanderthal Life, Love, Death and Art', Bloomsbury, UK, 2020, P.374
[44] Diversity-dependent speciation and extinction in hominins | Nature Ecology & Evolution
[45] https://www.theguardian.com/science/2012/sep/05/genes-genome-junk-dna-encode
[46] https://www.theguardian.com/science/2012/sep/05/genes-genome-junk-dna-encode
[47] Nessa Carey, 'Junk DNA – A Journey through the Dark Matter of the Genome', Icon Books Ltd, 2015, P.12
[48] https://en.wikipedia.org/wiki/Epigenetics
[49] Nessa Carey, 'The Epigenetics Revolution: How Modern Biology is Rewriting our Understanding of Genetics, Disease and Inheritance', Icon Books, 2011. P.2
[50] Nessa Care, 'The Epigenetics Revolution: How Modern Biology is Rewriting our Understanding of Genetics, Disease and Inheritance', Icon Books, 2011. P.8
[51] Nessa Care, 'The Epigenetics Revolution: How Modern Biology is Rewriting our Understanding of Genetics, Disease and Inheritance', Icon Books, 2011. P.8
[52] Nessa Care, 'The Epigenetics Revolution: How Modern Biology is Rewriting our Understanding of Genetics, Disease and Inheritance', Icon Books, 2011. P.9
[53] Fogel, Owens, Walsh, 'Artificial Intelligence through Simulated Evolution', John Wiley & Sons October 1st 1966
[54] Ingo Rechenberg (1971): Evolutions strategie - Optimierung technischer Systeme nach Prinzipien der biologischen Evolution (PhD thesis). Reprinted by Fromman-Holzboog (1973).
[55] Nessa Carey, 'The Epigenetics Revolution: How Modern Biology is Rewriting our Understanding of Genetics, Disease and Inheritance', Icon Books, 2011. P.309
[56] Nessa Carey, 'The Epigenetics Revolution: How Modern Biology is Rewriting our Understanding of Genetics, Disease and Inheritance', Icon Books, 2011. P.311
[57] How humans evolved to be both shockingly violent and super-cooperative | New Scientist
[58] How humans evolved to be both shockingly violent and super-cooperative | New Scientist
[59] https://en.wikipedia.org/wiki/Sociobiology
[60] https://en.wikipedia.org/wiki/Evolutionary_psychology
[61] Norman M Naimark, 'Genocide: A World History, Oxford University Press, USA, 2007, P.1
[62] Norman M Naimark, 'Genocide: A World History, Oxford University Press, USA, 2007, P.7
[63] Norman M Naimark, 'Genocide: A World History, Oxford University Press, USA, 2007, P.7
[64] 'The Armenian Genocide', Captivating History, 2019, P.48
[65] Ian Kershaw, 'Hitler', Penguin, UK, 2007, P.528,
[66] Jared Diamond, 'The Rise And Fall Of The Third Chimpanzee', 1991, Vintage Books, London UK, location 151
[67] Jared Diamond, 'The Rise And Fall Of The Third Chimpanzee', 1991, Vintage Books, London UK, location 162
[68] Finlayson, Clive, 'The Humans Who Went Extinct – Why Neanderthals died out and we survived', 2011, Clays Ltd, P.43
[69] https://www.wri.org/insights/2023-ipcc-ar6-synthesis-report-climate-change-findings, P.2
[70] Adam Kirsch, 'The Revolt Against Humanity – Imagining a Future Without Us', Columbia Global Reports, NY, 2023. P.20
[71] Finlayson, Clive, 'The Humans Who Went Extinct – Why Neanderthals died out and we survived', 2011, Clays Ltd, P.43
[72] Lovelock, James, 'The Vanishing Face of Gaia', Penguin Books Ltd. Kindle Edition, 2009, P.246.
[73] Hooper. Rowan, 'How the microbiome changes our idea of what it means to be human', New Scientist, October 2023
[74] About | CRC 1182 (metaorganism-research.com)
[75] Hooper. Rowan, 'How the microbiome changes our idea of what it means to be human', New Scientist, October 2023
[76] Microbiomes as sources of emergent host phenotypes | Science
[77] Tony Ord, 'The Precipice – Existential Risk and the Future of Humanity', Bloomsbury Publishing, 2020, London, P.206
[78] Tony Ord, 'The Precipice – Existential Risk and the Future of Humanity', Bloomsbury Publishing, 2020, London, P.206

[79] Charles Darwin, 'On the Origin of Species by Means of Natural Selection or the Preservation of Favoured Races in the Struggle for Life', United Edition Club, First Printing 1869 then reprinted by Heritage Press NY 1963
[80] Richard Hawkins, 'The Extended Phenotype – The Long Reach of the Gene', Oxford University Press, 1982,199, 2008, 2016, P.276
[81] Richard Hawkins, 'The Selfish Gene', Oxford University Press, 1976,1989, 2006, P.11
[82] Gerald M Elderman and Giulio Tononi, 'Consciousness – How Matter Becomes Imagination', The Penguin Press, GB, 2000, p.37
[83] Gerald M Elderman and Giulio Tononi, 'Consciousness – How Matter Becomes Imagination', The Penguin Press, GB, 2000, p.38
[84] Gerald M Elderman and Giulio Tononi, 'Consciousness – How Matter Becomes Imagination', The Penguin Press, GB, 2000, p.47
[85] Gerald M Elderman and Giulio Tononi, 'Consciousness – How Matter Becomes Imagination', The Penguin Press, GB, 2000, p.47
[86] The Mind as the Software of the Brain" by Ned Block, 1995, http://www.nyu.edu/gsas/dept/philo faculty/block/papers/msb.html, P.12
[87] Adam Pautz & Daniel Stoljar, 'Blockheads! Essays on Ned Block's Philosophy of Mind and Consciousness', MIT Press, Massachusetts USA, 2019
[88] Mathew Cobb, Why your brain is not a Computer', An edited extract from the book 'The Idea of the Brain, Profile, UK, 2020. Edited extract from The Guardian February 2020, https://www.theguardian.com/science/2020/feb/27/why-your-brain-is-not-a-computer-neuroscience-neural-networks-consciousness, p.1
[89] Mathew Cobb, Why your brain is not a Computer', An edited extract from the book 'The Idea of the Brain', Profile, UK, 2020. Edited extract from The Guardian February 2020, https://www.theguardian.com/science/2020/feb/27/why-your-brain-is-not-a-computer-neuroscience-neural-networks-consciousness, p.1
[90] https://www.humanbrainproject.eu/en/
[91] https://www.internationalbraininitiative.org/
[92] https://www.humanbrainproject.eu/en/silicon-brains/
[93] Jeff Hawkins' A Thousand Brains – A New Theory of Intelligence', Basic Books NY, 2021, P.4
[94] Jeff Hawkins' A Thousand Brains – A New Theory of Intelligence', Basic Books NY, 2021, P.99
[95] Hoel Erik P, Albantakis Larissa, & Tononi Giulio 'Quantifying causal emergence shows that macro can bit micro', Dept of Psychiatry, University of Wisconsin, Madison, WI 53719, 2013 http://www.pnas.org/cgi/doi/10.1073/pnas.1314922110 p 19794
[96] Sacks, Oliver, 'The River of Consciousness', Picador, 2017, P.104
[97] Sacks, Oliver, 'The River of Consciousness', Picador, 2017, P.106
[98] Clark. Andy, 'The Experience Machine – How our minds predict and shape reality', Allen Lane, USA, 2023
[99] Clark. Andy, 'The Experience Machine – How our minds predict and shape reality', Allen Lane, USA, 2023, P.5
[100] Clark. Andy, 'The Experience Machine – How our minds predict and shape reality', Allen Lane, USA, 2023, P.8
[101] Jeff Hawkins' A Thousand Brains – A New Theory of Intelligence', Basic Books NY, 2021, P.9
[102] Jeff Hawkins' A Thousand Brains – A New Theory of Intelligence', Basic Books NY, 2021, P.47
[103] John Noble Wilford, 'For the Sumerians, the writing may well be on the wall', New York Times, 16/4/02
[104] https://www.newscientist.com/article/mg25634202-800-mysterious-symbols-in-cave-paintings may-be-earliest-form-of-writing/
[105] https://lithub.com/so-gutenberg-didnt-actually-invent-the-printing-press/
[106] https://dictionary.cambridge.org/dictionary/english/intelligence
[107] https://www.collinsdictionary.com/dictionary/english/intelligence
[108] R. Yuste & M. Levin 'New Clues about the Origins of Biological Intelligence', , Scientific American, March/April 2022
[109] Peter Godfrey Smith, 'Other Minds: The Octopus and the Evolution of Intelligent Life, Harper Collins, 2017, P.54
[110] Peter Godfrey Smith, 'Other Minds: The Octopus and the Evolution of Intelligent Life, Harper Collins, 2017, P.68
[111] Peter Godfrey Smith, 'Other Minds: The Octopus and the Evolution of Intelligent Life, Harper Collins, 2017, P.57
[112] Jennifer Ackerman, 'The Bird Way – a new look at how birds talk, work, play, parent and think, 'Scribe Publications, Australia, 2020, P.23
[113] Jennifer Ackerman, 'The Bird Way – a new look at how birds talk, work, play, parent and think, 'Scribe Publications, Australia, 2020, P.23
[114] Jennifer Ackerman, 'The Bird Way – a new look at how birds talk, work, play, parent and think, 'Scribe Publications, Australia, 2020, P.24

[115] Jennifer Ackerman, 'The Bird Way – a new look at how birds talk, work, play, parent and think, 'Scribe Publications, Australia, 2020, P.25
[116] Jennifer Ackerman, 'The Bird Way – a new look at how birds talk, work, play, parent and think, 'Scribe Publications, Australia, 2020, P.31
[117] Jennifer Ackerman, 'The Bird Way – a new look at how birds talk, work, play, parent and think, 'Scribe Publications, Australia, 2020, P.31
[118] Jennifer Ackerman, 'The Bird Way – a new look at how birds talk, work, play, parent and think, 'Scribe Publications, Australia, 2020, P.32
[119] Wendy Williams, 'The Language of Butterflies" Black Inc, Australia, 2020, p148
[120] Wendy Williams, 'The Language of Butterflies" Black Inc, Australia, 2020, p148
[121] Wendy Williams, 'The Language of Butterflies" Black Inc, Australia, 2020, p149
[122] Wendy Williams, 'The Language of Butterflies" Black Inc, Australia, 2020, p148
[123] Sverdrup-Thygeson, 'Extraordinary Insects', Mudlark, an imprint of Harper Collins Publishers, 2019, P.xv
[124] Sverdrup-Thygeson, 'Extraordinary Insects', Mudlark, an imprint of Harper Collins Publishers, 2019, P.xviii
[125] Sverdrup-Thygeson, 'Extraordinary Insects', Mudlark, an imprint of Harper Collins Publishers, 2019, P.24
[126] Sverdrup-Thygeson, 'Extraordinary Insects', Mudlark, an imprint of Harper Collins Publishers, 2019, P.28
[127] Chittka.Lars, 'The Mind of a Bee', Princeton University Press, USA, 2022, P.71
[128] Chittka.Lars, 'The Mind of a Bee', Princeton University Press, USA, 2022, P.77.
[129] Chittka.Lars, 'The Mind of a Bee', Princeton University Press, USA, 2022, P.49
[130] Chittka.Lars, 'The Mind of a Bee', Princeton University Press, USA, 2022, P.208
[131] Anil Seth, 'Being You – A New Science of Consciousness', Faber Books, UK, 2021, P.384
[132] Masson. Jeffrey, 'The Secret World of Farm Animals', Penguin Random House, UK, 2004, P.137
[133] Stevens S S, (1973) Edwin Garrigues Boring, National Academy of Sciences, 38-76,
[134] Malcolm Gladwell, Outliers, The Story of Success, Penguin Books, 2017, P.76
[135] Malcolm Gladwell, Outliers, The Story of Success, Penguin Books, 2017, P.299
[136] Malcolm Gladwell, Outliers, The Story of Success, Penguin Books, 2017, P.78
[137] https://dictionary.cambridge.org/dictionary/english/emotional-intelligence
[138] Russell M Kendrick, 'Understanding Your Emotional Intelligence', Kindle Edition, P.2
[139] Daniel Goleman, 'Emotional Intelligence', Bloomsbury Publishing, Kindle Edition, P.44
[140] Russell M Kendrick, 'Understanding Your Emotional Intelligence', Kindle Edition, P.2
[141] https://dictionary.cambridge.org/dictionary/english/sentience
[142] https://www.collinsdictionary.com/dictionary/english/sentient
[143] Nicholas Humphrey's 'Sentience', Oxford University Press< UK, 2022, P.xii
[144] Nicholas Humphrey's 'Sentience', Oxford University Press< UK, 2022, P.x11
[145] https://www.youtube.com/watch?v=C5RiHTSXK2A
[146] https://www.youtube.com/watch?v=gna3PU81xSw
[147] Peter Godfrey Smith, 'Other Minds: The Octopus and the Evolution of Intelligent Life, Harper Collins, 2017, P.62
[148] Chittka, Lars, 'The Mind of a Bee', Princeton University Press, new Jersey USA, P.192
[149] Jonathan Birch, 'The philosophy of Social Evolution', Oxford University Press, UK, 2017, P.3
[150] Sacks, Oliver, 'The River of Consciousness', Picador, 2017, P.50
[151] Sacks, Oliver, 'The River of Consciousness', Picador, 2017, P.52
[152] https://www.theguardian.com/commentisfree/2021/may/16/animals-feel-humans-evidence sentient
[153] Yong. Ed, 'An Immense World – How Animal Senses Reveal the Hidden Realms Around Us', Vintage, 2022, P.12
[154] Nagel, Thomas (1974). 'What is It Like to Be a Bat? The Philosophical Review. 83 (4): P.435–450
[155] https://www.theguardian.com/commentisfree/2021/may/16/animals-feel-humans-evidence sentient
[156] https://dictionary.cambridge.org/dictionary/english/consciousness
[157] https://www.collinsdictionary.com/dictionary/english/consciousness
[158] Nicholas Humphrey's 'Sentience', Oxford University Press, UK, 2022, P.4
[159] https://dictionary.cambridge.org/dictionary/english/awareness
[160] Anil Seth, 'Being You – A New Science of Consciousness', Faber Books, UK, 2021, P.24
[161] Nicholas Humphrey's 'Sentience', Oxford University Press< UK, 2022, P.113
[162] Nicholas Humphrey's 'Sentience', Oxford University Press< UK, 2022, P.202
[163] Peter Godfred-Smith, Metazoa – Animal Minds and the Birth of Consciousness', William Collins, London 2020, loc 1815 of 5173
[164] Nicholas Humphrey's 'Sentience', Oxford University Press< UK, 2022, P.206
[165] The New York Declaration on Animal Consciousness (google.com)
[166] Anil Seth, 'Being You – A New Science of Consciousness', Faber Books, UK, 2021

[167] Anil Seth, 'Being You – A New Science of Consciousness', Faber Books, UK, 2021, P.15
[168] Anil Seth, 'Being You – A New Science of Consciousness', Faber Books, UK, 2021, P16
[169] Anil Seth, 'Being You – A New Science of Consciousness', Faber Books, UK, 2021, P21
[170] https://consc.net/
[171] https://consc.net/papers/universal.pdf
[172] Birch, J., Schnell, A. K., & Clayton, N. S. (2020). 'Dimensions of animal consciousness'. Trends in Cognitive Sciences,24 (10), 789–801.
[173] Birch, J., Schnell, A. K., & Clayton, N. S. (2020). 'Dimensions of animal consciousness'. Trends in Cognitive Sciences, 24 (10), 789
[174] Bayne. T., Hohwy. J., Owen. A.; 'Are There Level of Consciousness?', Trends in Cognitive Sciences, June 2016, Vol. 20, No. 6, P.406
[175] Birch, J., Schnell, A. K., & Clayton, N. S. (2020). 'Dimensions of animal consciousness'. Trends in Cognitive Sciences,24 (10), 789
[176] https://www.nhm.ac.uk/discover/octopuses-keep-surprising-us-here-are-eight-examples-how.html
[177] merlin sheldrake, 'Entangled Life – How Fungi Makes our Worlds, Change Our Minds, and Shape Our futures'Vintage , partI
[178] Suzanne Simard, 'Finding the Mother Tree – Uncovering the Wisdom and Intelligence of the Forest', Penguin 2021
[179] Peter Wohlleben in 'The Hidden Life of Trees', Black Inc, an imprint of Schwartz Publishing, 2016, P.160
[180] Suzanne Simard, 'Finding the Mother Tree – Uncovering the Wisdom and Intelligence of the Forest', Penguin Books, Canada, 2021 p.4
[181] Suzanne Simard, 'Finding the Mother Tree – Uncovering the Wisdom and Intelligence of the Forest', Penguin Books, Canada, 2021 p.5
[182] Suzanne Simard, 'Finding the Mother Tree – Uncovering the Wisdom and Intelligence of the Forest', Penguin Books, Canada, 2021 p.45
[183] Peter Wohlleben in 'The Hidden Life of Trees', Black Inc, an imprint of Schwartz Publishing, 2016
[184] Peter Wohlleben in 'The Hidden Life of Trees', Black Inc, an imprint of Schwartz Publishing, 2016, P.7
[185] MacGowan Roger A, Ordway III, Frederick, 'Intelligence in the Universe, Prentice-Hall, USA, 1966, P.93
[186] https://www.vocabulary.com/dictionary/vanishing
[187] https://dictionary.cambridge.org/dictionary/english/machine
[188] https://dictionary.cambridge.org/dictionary/english/autonomy
[189] https://pediaa.com/
[190] https://pediaa.com/what-is-the-difference-between-autonomous-and-independent/
[191] O'Regan, Gerard (2010). A Brief History of Computing. Springer Nature. p. 65 ISBN 9783030665999
[192] Some Enigma machines collected from the Germans were given by the British to their African colonies after the war to help with 'secure' communications in these countries and were in use up to the early 1960's. Interest in reading another country's internal documents sounds addictive.
[193] Turing, A. M. (1937). "On Computable Numbers, with an Application to the Entscheidungs problem". Proceedings of the London Mathematical Society. 2. 42 (1): 230–265
[194] https://thenextweb.com/news/20-years-ago-today-the-world-wide-web-opened-to-the-public
[195] https://www.statista.com/statistics/617136/digital-population-worldwide/
[196] https://www.dictionary.com/browse/symbiotic
[197] Turing A.M 'Computing Machinery and intelligence', Mind 49, 1950, p.433-460. Special note of a link to the University of Mayland, a public research university, for the original paper available on line. https://redirect.cs.umbc.edu/courses/471/papers/turing.pdf
[198] https://www. mdb.com/title/tt2084970/
[199] Belden, Thomas; Belden, Marva 'The lengthening Shadow" The Life of Thomas J Watson, Little, Brown and Company, 1962, P.157
[200] Belden, Thomas; Belden, Marva 'The lengthening Shadow" The Life of Thomas J Watson, Little, Brown and Company, 1962, P.158
[201] https://geekhistory.com/content/urban-legend-i-think-there-world-market-maybe-five-computers
[202] Rodwin, Noah, 'Time Berners-Lee and the World Wide Web' Kindle Book, location 2 of 130
[203] Rodwin, Noah, 'Time Berners-Lee and the World Wide Web' Kindle Book, location 30 of 130
[204] A. M Turing Award, Association for Computing Machinery 2016. https://amturing.acm.org/award winners/berners-lee_8087960.cfm
[205] Gradwell. Malcolm, 'Outliers – The Story of Success', Penguin Books, USA, 2008, P.51
[206] Gradwell. Malcolm, 'Outliers – The Story of Success', Penguin Books, USA, 2008, P.49
[207] Gradwell. Malcolm, 'Outliers – The Story of Success', Penguin Books, USA, 2008, P.49
[208] Gradwell. Malcolm, 'Outliers – The Story of Success', Penguin Books, USA, 2008, P.49
[209] Gradwell. Malcolm, 'Outliers – The Story of Success', Penguin Books, USA, 2008, P.47
[210] Manes, Stephen & Andrews, Paul, 'Gates', Cadwallader & Stern, 1993 – 20113, Location 374 of 13485
[211] https://edition.cnn.com/2021/08/02/tech/bill-melinda-gates-divorce-finalized/index.html

[212] https://companiesmarketcap.com/microsoft/revenue/
[213] https://zurb.com/blog/steve-jobs-and-xerox-the-truth-about-inno
[214] https://www.businessinsider.com/malcolm-gladwell-on-steve-jobs-2014-10
[215] Stone, Brad. 'Amazon Unbound', Simon & Schuster, NY, 2021, P.4
[216] Stone, Brad. 'Amazon Unbound', Simon & Schuster, NY, 2021, P.5
[217] Stone, Brad. 'Amazon Unbound', Simon & Schuster, NY, 2021, P.5
[218] Stone, Brad. 'Amazon Unbound', Simon & Schuster, NY, 2021, P.7
[219] Stone, Brad. 'Amazon Unbound', Simon & Schuster, NY, 2021, P.8
[220] Stone, Brad. 'Amazon Unbound', Simon & Schuster, NY, 2021, P.9
[221] https://www.sec.gov/ix?doc=/Archives/edgar/data/1018724/000101872423000004/amzn-20221231 htm, P.23
[222] https://economictimes.indiatimes.com/markets/stocks/news/a-16k-investment-on-google-in-2004 ipo-would-be-worth-1-million-today-heres-how/tech-titan/slideshow/103530251.cms
[223] https://au.pcmag.com/ai/103683/meta-apps-hit-31-billion-daily-users-zuckerberg-touts-open source-ai-tools
[224] https://www.businessinsider.com/meta-lost-30-billion-on-metaverse-rivals-spent-far-less-2022 10?op=1
[225] https://www.businessinsider.com/meta-lost-30-billion-on-metaverse-rivals-spent-far-less-2022 10?op=1#apples-iphone-1
[226] https://marketrealist.com/who-owns-netflix-the-184-90-billion-global-streaming-company/
[227] https://venturebeat.com/media/global-video-streaming-market-is-largely-controlled-by-the-usual suspects/
[228] https://marketrealist.com/who-owns-netflix-the-184-90-billion-global-streaming-company/
[229] https://www.rankdex.com/about.html
[230] https://www.rankdex.com/about.html
[231] https://companiesmarketcap.com/baidu/marketcap/
[232] Hu. Leng, 'Ma Huateng & Tencent – The Story of an Online Chinese Empire', LID Publishing, 2017, P.11
[233] Hu. Leng, 'Ma Huateng & Tencent – The Story of an Online Chinese Empire', LID Publishing, 2017, P.24
[234] Renard. Philip, 'Tencent: The Octopus that Rules the World', Kindle Edition, 2023, P.9
[235] Renard. Philip, 'Tencent: The Octopus that Rules the World', Kindle Edition, 2023, P.11
[236] Hu. Leng, 'Ma Huateng & Tencent – The Story of an Online Chinese Empire', LID Publishing, 2017, P.27
[237] Renard. Philip, 'Tencent: The Octopus that Rules the World', Kindle Edition, 2023, P.13
[238] Renard. Philip, 'Tencent: The Octopus that Rules the World', Kindle Edition, 2023, P.18
[239] Renard. Philip, 'Tencent: The Octopus that Rules the World', Kindle Edition, 2023, P.24
[240] Renard. Philip, 'Tencent: The Octopus that Rules the World', Kindle Edition, 2023, P.29
[241] Renard. Philip, 'Tencent: The Octopus that Rules the World', Kindle Edition, 2023, P.32
[242] Hu. Leng, 'Ma Huateng & Tencent – The Story of an Online Chinese Empire', LID Publishing, 2017, P.68
[243] Renard. Philip, 'Tencent: The Octopus that Rules the World', Kindle Edition, 2023, P.32
[244] Renard. Philip, 'Tencent: The Octopus that Rules the World', Kindle Edition, 2023, P.33
[245] Renard. Philip, 'Tencent: The Octopus that Rules the World', Kindle Edition, 2023, P.38
[246] Renard. Philip, 'Tencent: The Octopus that Rules the World', Kindle Edition, 2023, P.38
[247] Clark. Duncan, 'Alibaba – The House that Jack Ma Built', CCC, An Imprint of Harper Collins, USA, 2016, P.37
[248] Clark. Duncan, 'Alibaba – The House that Jack Ma Built', CCC, An Imprint of Harper Collins, USA, 2016, P.66
[249] Clark. Duncan, 'Alibaba – The House that Jack Ma Built', CCC, An Imprint of Harper Collins, USA, 2016, P.75
[250] Clark. Duncan, 'Alibaba – The House that Jack Ma Built', CCC, An Imprint of Harper Collins, USA, 2016, P.146
[251] Clark. Duncan, 'Alibaba – The House that Jack Ma Built', CCC, An Imprint of Harper Collins, USA, 2016, P.175
[252] Clark. Duncan, 'Alibaba – The House that Jack Ma Built', CCC, An Imprint of Harper Collins, USA, 2016, P.179
[253] https://technode.com/2018/01/25/alibaba-500-million-market-cap/
[254] Isaacson. Walter, 'Elon Musk', Simon & Schuster, Kindle Edition, USA, 2023, P.35
[255] Isaacson. Walter, 'Elon Musk', Simon & Schuster, Kindle Edition, USA, 2023, P.7
[256] Isaacson. Walter, 'Elon Musk', Simon & Schuster, Kindle Edition, USA, 2023, P.72
[257] Isaacson. Walter, 'Elon Musk', Simon & Schuster, Kindle Edition, USA, 2023, P.116
[258] http://news.bbc.co.uk/2/hi/technology/8120324.stm
[259] Simon, Mark, 'TikTok', independently published, USA, 2020, P.1
[260] https://www.bbc.com/news/business-41939296

[261] King. Herbert, 'Let's TikTok – The Beginners Guide for your TikTok Success' Kindle online, US, 2022, P.5
[262] Hart. Doris J, 'TikTok Ads Unleased', Kinde Store, 2024, P.4
[263] King. Herbert, 'Let's TikTok – The Beginners Guide for your TikTok Success'. Kindle online, US, 2022, P.25
[264] https://www.nbcnews.com/tech/tech-news/tiktok-surpasses-google-popular-website-year-new data-suggests-rcna9648
[265] https://www.businessinsider.com/tiktok-billionaire-zhang-yiming-net-worth-lifestyle-2019 11#zhang-founded-tiktoks-parent-company-in-2012-4
[266] https://www.bloomberg.com/news/articles/2021-03-30/bytedance-is-said-valued-at-250-billion-in private-trades
[267] https://www.cnet.com/tech/mobile/apple-music-vs-apple-podcast-vs-apple-tv-whats-the difference/
[268] https://arstechnica.com/gadgets/2023/01/apple-previews-a-trio-of-apps-that-will-finally-replace itunes-for-windows/
[269] https://www.pewresearch.org/short-reads/2023/11/22/online-shopping-has-grown-rapidly-in-u-s but-most-sales-are-still-in-stores/
[270] https://www.ecns.cn/news/economy/2024-01-19/detail-ihcwvwfu7684762.shtml
[271] https://www.reuters.com/technology/chatgpt-sets-record-fastest-growing-user-base-analyst note-2023-02-01/
[272] Wiener, Norbert (1948). Cybernetics: Or Control and Communication in the Animal and the Machine. Cambridge, Massachusetts: MIT Press.
[273] Wiener, Norbert (1948). Cybernetics: Or Control and Communication in the Animal and the Machine. Cambridge, Massachusetts: MIT Press, P. xviii
[274] Pere Dubarle, http://research.mchrbn.net/ref/2013/02/1948-machines-a-gouverner/ review of Wiener book Cybernetics
[275] John Brockman, 'Possible Minds - 25 Ways of Looking at AI', Penguin Books, 2020, P xix
[276] John Brockman, 'Possible Minds - 25 Ways of Looking at AI', Penguin Books, 2020, P xix
[277] McCarthy along with Shannon, Minsky & Rochester proposed the term 'artificial intelligence' in a paper to the Dartmouth Conference in 1956
[278] Fouad Sabry, 'Artificial General Intelligence',www.1bkofficial.org, P.41
[279] Hans Weber, 'Artificial Intelligence and Life', P.1
[280] Gary Marcus and Ernest David, 'Rebooting AI – Building Artificial We Can Trust', Pantheon Books New York USA 2019, location 773 of 5057
[281] Fouad Sabry, Artificial General Intelligence, 2022, P.43
[282] Fouad Sabry, Artificial General Intelligence, 2022, P.43
[283] Fouad Sabry, Artificial General Intelligence, 2022, P.53
[284] Hines. Harwood C, 'From Code to Consciousness: Exploring the Depths of AI', online publication, 2023, P.24
[285] Hines. Harwood C, 'From Code to Consciousness: Exploring the Depths of AI', online publication, 2023, P.24
[286] https://www.zmescience.com/future/chatgpt-launched-just-a-week-ago-and-rocked-the-world heres-what-you-need-to-know-about-it/
[287] https://www.cnbc.com/2022/12/13/chatgpt-is-a-new-ai-chatbot-that-can-answer-questions-and write-essays.html
[288] https://fortune.com/2023/01/23/microsoft-investing-10-billion-open-ai-chatgpt/
[289] https://www.forbes.com/sites/ariannajohnson/2022/12/07/heres-what-to-know-about-openais chatgpt-what-its-disrupting-and-how-to-use-it/?sh=454521e92643
[290] Brian Christian, 'The Alignment Problem – How can Machines Learn Human Values', Atlantic Books, USA, 2020, P.25
[291] Brian Christian, 'The Alignment Problem – How can Machines Learn Human Values', Atlantic Books, USA, 2020, P.388
[292] https://www.merriam-webster.com/dictionary/artificial%20intelligence
[293] Ejeke. Patrick, 'Web 3.0 – What is Web3?' Patrick Ejeke, 2022, P.7
[294] Ejeke. Patrick, 'Web 3.0 – What is Web3?' Patrick Ejeke, 2022, P.3
[295] https://dictionary.cambridge.org/dictionary/english/bot
[296] https://en.wikipedia.org/wiki/Internet_bot
[297] Zeifman, Igal (January 24, 2017). "Bot Traffic Report 2016". Incapsula. Retrieved February 1, 2017
[298] https://www.imperva.com/blog/bot-traffic-report-2016/?redirect=Incapsula
[299] https://en.wikipedia.org/wiki/Internet_bot
[300] Howard, Philip N (October 18, 2018). *"How Political Campaigns Weaponize Social Media Bots"*. IEEE Spectrum
[301] The Mind as the Software of the Brain" by Ned Block, 1995, http://www.nyu.edu/gsas/dept/philo faculty/block/papers/msb.html, P.3

[302] https://www.youtube.com/watch?v=7qnd-hdmgfk
[303] https://web.archive.org/web/20170306034914/http://www businesswire.com/news home/20100217005514/en/Pacifcor-Names-Latham-Watkins-Field-Terminator-Inquiries
[304] https://www.windowscentral.com/history-cortana-microsofts-digital-assistant
[305] https://www.zdnet.com/article/whatever-happened-to-cortana/
[306] https://www.theguardian.com/technology/2022/aug/09/blenderbot-meta-chatbot-facebook
[307] https://ai.facebook.com/blog/state-of-the-art-open-source-chatbot/
[308] https://onlinelibrary.wiley.com/doi/abs/10.1002/mar.21480
[309] https://www.brookings.edu/research/how-ai-bots-and-voice-assistants-reinforce-gender-bias/
[310] https://www.theverge.com/2021/7/21/22587130/alexa-masculine-voice-gender-ziggy-amazon smart-assistant
[311] https://joernlengsfeld.com/en/definition/digital-life/
[312] https://www.ipxo.com/blog/how-big-is-the-internet/ author Beatričė Raščiūtė
[313] https://www.ipxo.com/blog/how-big-is-the-internet/ author Beatričė Raščiūtė
[314] Taylor. Allen, 'Web3 Social: How Creators Are Changing the World Wide Web (And You Can Too!)', Allen Taylor, 2023, location 76
[315] Taylor. Allen, 'Web3 Social: How Creators Are Changing the World Wide Web (And You Can Too!)', Allen Taylor, 2023, location 90
[316] Taylor. Allen, 'Web3 Social: How Creators Are Changing the World Wide Web (And You Can Too!)', Allen Taylor, 2023, location 76
[317] Taylor. Allen, 'Web3 Social: How Creators Are Changing the World Wide Web (And You Can Too!)', Allen Taylor, 2023, location 996
[318] Accialini. Nicola, 'Web3 Revolution – Blockchain, Cryptocurrencies, NFT and Metaverse', Accialini Consulting, 2022, location 123
[319] Taylor. Allen, 'Web3 Social: How Creators Are Changing the World Wide Web (And You Can Too!)', Allen Taylor, 2023, location 90
[320] Ejeke. Patrick, 'Web 3.0 – What is Web3?' Patrick Ejeke, 2022, P.5
[321] https://solidproject.org/about
[322] Ejeke. Patrick, 'Web 3.0 – What is Web3?' Patrick Ejeke, 2022, P.9
[323] Ejeke. Patrick, 'Web 3.0 – What is Web3?' Patrick Ejeke, 2022, P.9
[324] 'We've discovered the secret of immortality. The bad news is it's not for us': why the godfather of AI fears for humanity | Artificial intelligence (AI) | The Guardian Article by Alex Hern, Technology reporter for the Guardian May 2023
[325] https://dictionary.cambridge.org/dictionary/english/intelligence
[326] R. Yuste & M. Levin 'New Clues about the Origins of Biological Intelligence', , Scientific American, March/April 2022
[327] Acknowledgement here to the Cambridge Dictionary and Yuste & Levin for phrases that, when combined, gave a new definition.
[328] Lepore. Jill, 'IF THEN – How one data company invented the Future', 2020, John Murray Publishers, Great Britain, P.12
[329] https://www.researchgate.net/publication/2928775_The_Mind_as_the_Software_of_the_Brain
[330] The Mind as the Software of the Brain" by Ned Block, 1995, http://www.nyu.edu/gsas/dept/philo faculty/block/papers/msb.html, P.2
[331] https://dictionary.cambridge.org/dictionary/english/sentience
[332] https://www.collinsdictionary.com/dictionary/english/sentient
[333] Tao, Jianhua; Tieniu Tan (2005). "Affective Computing: A Review". Affective Computing and Intelligent Interaction. Vol. LNCS 3784. Springer. pp. 981–995
[334] https://dictionary.cambridge.org/dictionary/english/affective
[335] https://www.media.mit.edu/people/picard/overview/
[336] Picard, Rosalind (1997). Affective Computing. Cambridge, MA: MIT Press. p. 1.
[337] Ricard. Rosalind W, (2000), 'Affective Computing', MIT Press, Cambridge, MA
[338] Ricard. Rosalind W, (2000), 'Affective Computing', MIT Press, Cambridge, MA, P.249
[339] Ricard. Rosalind W, (2000), 'Affective Computing', MIT Press, Cambridge, MA. P. 250
[340] Ricard. Rosalind W, (2000), 'Affective Computing', MIT Press, Cambridge, MA, P.250
[341] Ricard. Rosalind W, (2000), 'Affective Computing', MIT Press, Cambridge, MA, P.250
[342] Ricard. Rosalind W, (2000), 'Affective Computing', MIT Press, Cambridge, MA, P.252
[343] https://www.youtube.com/watch?v=DQacCB9tDaw 16/05/24
[344] https://www.newscientist.com/article/2323905-has-googles-lamda-artificial-intelligence-really achieved-sentience/
[345] Is LaMDA Sentient? — an Interview | by Blake Lemoine | Medium
[346] Is LaMDA Sentient? — an Interview | by Blake Lemoine | Medium
[347] Turing experts respond to Google AI Chatbot story in the media | The Alan Turing Institute
[348] Nonsense on Stilts - by Gary Marcus - Marcus on AI (substack.com)
[349] Sentience and AI: A Dialog between Gary Marcus and Blake Lemoine (substack.com)
[350] Sentience and AI: A Dialog between Gary Marcus and Blake Lemoine (substack.com)

351 Nonsense on Stilts - by Gary Marcus - Marcus on AI (substack.com)
352 https://www.collinsdictionary.com/dictionary/english/sentient
353 R. Yuste & M. Levin 'New Clues about the Origins of Biological Intelligence', , Scientific American, March/April 2022
354 https://www.newscientist.com/article/2323905-has-googles-lamda-artificial- ntelligence-really achieved-sentience/
355 https://en.wikipedia.org/wiki/Turing_test
356 Turing Test in Artificial Intelligence - GeeksforGeeks
357 https://dictionary.cambridge.org/dictionary/english/simulating
358 Ricard. Rosalind W, (2000), 'Affective Computing', MIT Press, Cambridge, MA P.250
359 https://www.newscientist.com/article/2323905-has-googles-lamda-artificial-intelligence-really achieved-sentience/
360 Stephen Thaler Claims He's Built a Sentient AI - The New Stack
361 Kirshner. Sam, Vidgen. Richard, Wallace. Catriona, (and others) 'Checkmate Humanity – The how and why or Responsible AI', Global Stories, Australia, 2022, P.23
362 Morris. Gregory, 'Human, Obsolete – How AI Will Upend Human Life As We Know It', Independently Published, 2023, P.16
363 Chittka, Lars, 'The Mind of a Bee', Princeton University Press, new Jersey USA, P.207
364 https://dictionary.cambridge.org/dictionary/english/consciousness
365 Humphries, Nicholas, 'Sentience', Oxford University Press, UK, 2022, P.4
366 Nicholas Humphrey's 'Sentience', Oxford University Press< UK, 2022, P.113
367 Schneider. Susan, 'Artificial You – AI and the future of your Mind', Princeton University Press, 2019, P.16
368 Schneider. Susan, 'Artificial You – AI and the future of your Mind', Princeton University Press, 2019, P.17
369 https://consc.net/papers/universal.pdf
370 Schneider. Susan, 'Artificial You – AI and the future of your Mind', Princeton University Press, 2019, P.17
371 Schneider. Susan, 'Artificial You – AI and the future of your Mind', Princeton University Press, 2019, P.18
372 https://science.ku.dk/english/press/news/2023/jellyfish-are-smarter-than-you think/#:~:text=Jellyfish%20are%20more%20advanced%20than,cells%20and%20no%20 centralized%20brain.
373 MacGowan Roger A, Ordway III, Frederick, 'Intelligence in the Universe, Prentice-Hall, USA, 1966, P.92
374 MacGowan Roger A, Ordway III, Frederick, 'Intelligence in the Universe, Prentice-Hall, USA, 1966, P.93
375 MacGowan Roger A, Ordway III, Frederick, 'Intelligence in the Universe, Prentice-Hall, USA, 1966, P.92
376 MacGowan Roger A, Ordway III, Frederick, 'Intelligence in the Universe, Prentice-Hall, USA, 1966, P.93
377 MacGowan Roger A, Ordway III, Frederick, 'Intelligence in the Universe, Prentice-Hall, USA, 1966, P.93
378 https://dictionary.cambridge.org/dictionary/english/singularity#
379 http://www.stat.vt.edu/tech_reports/2005/GoodTechReport.pdf
380 Vinge. Vernor, 'The Coming Technological Singularity: How to Survive in the Post-Human Era', NASA Vision-21 Symposium, March 3021, 1993
381 Kurzweil. Ray,' The singularity is Near: When Humans Transcend Biology', Duckworth Overlook, UK, 2005, P.28
382 Kurzweil. Ray,' The singularity is Near: When Humans Transcend Biology', Duckworth Overlook, UK, 2005, P.30
383 Kurzweil. Ray,' The singularity is Near: When Humans Transcend Biology', Duckworth Overlook, UK, 2005, P.190
384 Kurzweil. Ray,' The singularity is Near: When Humans Transcend Biology', Duckworth Overlook, UK, 2005, P430
385 Kurzweil. Ray,' The singularity is Near: When Humans Transcend Biology', Duckworth Overlook, UK, 2005, P442
386 Collection of sources defining "singularity" singularitysymposium.com. 17 Apr l 2019.
387 Ford. Martin, 'The Rise of the Robots – Technology and the Threat of Mass Unemployment', Basic Books, USA, 2015
388 Ford. Martin, 'The Rise of the Robots – Technology and the Threat of Mass Unemployment', Basic Books, USA, 2015, P.228
389 Ford. Martin, 'The Rise of the Robots – Technology and the Threat of Mass Unemployment', Basic Books, USA, 2015, P.229

[390] Ford. Martin, 'The Rise of the Robots – Technology and the Threat of Mass Unemp0loyment', Basic Books, USA, 2015, P.235
[391] Chomsky, Noam: The Singularity is Science Fiction (interview), YouTube, October 4, 2013
[392] Hines. Harwood C, 'From Code to Consciousness: Exploring the Depths of AI', online publication, 2023, P.24
[393] Gradwell. Malcolm, 'Outliers – The Story of Success', Penguin Books, USA, 2008, P.74
[394] Gradwell. Malcolm, 'Outliers – The Story of Success', Penguin Books, USA, 2008, P.78
[395] Tony Ord, 'The Precipice – Existential Risk and the Future of Humanity', Bloomsbury Publishing, 2020, London, P.206
[396] Huxley. Julian, 'Transhumanism', reprinted in Mentor Books, USA under title 'Knowledge, Morality & Destiny', 1957, P.76
[397] Bess. Michael, 'Make Way for the Superhumans – How the science of bio-enhancement is transforming our world, and how we need to deal with it', Icon Books, USA, 2015, P.28
[398] Bess. Michael, 'Make Way for the Superhumans – How the science of bio-enhancement is transforming our world, and how we need to deal with it', Icon Books, USA, 2015, P.31
[399] Bess. Michael, 'Make Way for the Superhumans – How the science of bio-enhancement is transforming our world, and how we need to deal with it', Icon Books, USA, 2015, P.33
[400] Bess. Michael, 'Make Way for the Superhumans – How the science of bio-enhancement is transforming our world, and how we need to deal with it', Icon Books, USA, 2015, P.36
[401]Bess. Michael, 'Make Way for the Superhumans – How the science of bio-enhancement is transforming our world, and how we need to deal with it', Icon Books, USA, 2015, P.37
[402] **Thousands_of_AI_authors_on_the_future_of_AI.pdf (aiimpacts.org), P.20**
[403] Ford. Martin, 'The Rise of the Robots – Technology and the Threat of Mass Unemp0loyment', Basic Books, USA, 2015, P.212
[404] Ford. Martin, 'The Rise of the Robots – Technology and the Threat of Mass Unemp0loyment', Basic Books, USA, 2015, P.213
[405] Ford. Martin, 'The Rise of the Robots – Technology and the Threat of Mass Unemp0loyment', Basic Books, USA, 2015, P.217
[406] Smith. Noah, 'Drones will cause an Upheaval of Society like we haven't seen in 700 years', Quartz, March 11, 2014
[407] Crutzen, Paul J. and Stoermer, Eugene F.. ""The 'Anthropocene'" (2000)". *The Future of Nature: Documents of Global Change,* edited by Libby Robin, Sverker Sörlin and Paul Warde, New Haven: Yale University Press, 2013, pp. 479-490. **https://doi.org/10.12987/9780300188479-041**
[408] **https://www.smithsonianmag.com/science-nature/what-happened-worlds-most enormousanimals-180964255/**
[409] **https://public.oed.com/blog/june-2014-update-new-words-notes/**
[410] **Thousands_of_AI_authors_on_the_future_of_AI.pdf (aiimpacts.org), P.20**
[411] MacGowan Roger A, Ordway III, Frederick, 'Intelligence in the Universe, Prentice-Hall, USA, 1966, P.270
[412] MacGowan Roger A, Ordway III, Frederick, 'Intelligence in the Universe, Prentice-Hall, USA, 1966, P.360
[413] MacGowan Roger A, Ordway III, Frederick, 'Intelligence in the Universe, Prentice-Hall, USA, 1966, P.270-271
[414] MacGowan Roger A, Ordway III, Frederick, 'Intelligence in the Universe, Prentice-Hall, USA, 1966, P.271
[415] MacGowan Roger A, Ordway III, Frederick, 'Intelligence in the Universe, Prentice-Hall, USA, 1966, P.272
[416] MacGowan Roger A, Ordway III, Frederick, 'Intelligence in the Universe, Prentice-Hall, USA, 1966, P.232
[417] MacGowan Roger A, Ordway III, Frederick, 'Intelligence in the Universe, Prentice-Hall, USA, 1966, P.272
[418] MacGowan Roger A, Ordway III, Frederick, 'Intelligence in the Universe, Prentice-Hall, USA, 1966, P.272
[419] MacGowan Roger A, Ordway III, Frederick, 'Intelligence in the Universe, Prentice-Hall, USA, 1966, P.234
[420] Machinacene is a new word in this context, but 'Machinacene Epoch' is the name of an album by the USA band The FMs in 2017, and that is the first use of the term on a Google search **https://thefms.bandcamp.com/album/machinacene-epoch**
[421] Nessa Care, 'The Epigenetics Revolution: How Modern Biology is Rewriting our Understanding of Genetics, Disease and Inheritance', Icon Books, 2011. P.8
[422] **https://explodingtopics.com/blog/chatgpt-users#how-many June 2024**
[423] Nagel, Thomas (1974). 'What is It Like to Be a Bat? The Philosophical Review. 83 (4): P.435–450
[424] Wrangham.Richard and Peterson. Dale, 'Demonic Males – Apes and the Origins of Human Violence', Mariner Books, Harper Collins, NY USA, P.108-126. For an in-depth analysis of male violence, in the context of history, culture and gender, Chapter 6 is a great in-depth analysis.

Made in the USA
Monee, IL
04 November 2026

67380048R00128